Praise for (

"[Adam Rutherford's] scientific demolition of the eugenic project is brilliantly illuminating and compelling. His book will be indispensable for anyone who wants to assess the wild claims and counter-claims surrounding new genetic technologies."

—John Gray, *New Statesman*

"Discussions around the idea of population control are increasingly resurfacing. *Control*'s strength is that it provides not only much-needed guidance for these conversations by reminding us of the horrors of the past, but also uses scientific evidence to dismantle the viability of these ideas."

—Layal Liverpool, *New Scientist*

"Rutherford's swift, well-written account of these fascinating scientific and moral issues is well worth a read."

—Emma Duncan, *Times* (UK)

"Rutherford presents a profoundly sensible take on the complexities of history . . . an important book."

—Shaoni Bhattacharya, *Mail on Sunday*

"Fizzy and pugnacious . . . brilliant. . . . A fierce and funny broadside against eugenics and its admirers."

—Steven Poole, *Sunday Telegraph*

CONTROL

CONTROL

THE DARK HISTORY AND TROUBLING
PRESENT OF EUGENICS

ADAM RUTHERFORD

W. W. NORTON & COMPANY
Celebrating a Century of Independent Publishing

For information about permission to reproduce selections from this book, write to
Permissions, W. W. Norton & Company, Inc., 500 Fifth Avenue, New York, NY 10110

For information about special discounts for bulk purchases, please contact
W. W. Norton Special Sales at specialsales@wwnorton.com or 800-233-4830

Manufacturing by Lakeside Book Company
Book design by Chris Welch
Production manager: Lauren Abbate

Library of Congress Cataloging-in-Publication Data

Names: Rutherford, Adam, author.
Title: Control : the dark history and troubling present of eugenics / Adam Rutherford.
Description: First American edition. | New York, NY : W. W. Norton & Company,
[2023] | Includes bibliographical references and index.
Identifiers: LCCN 2022027417 | ISBN 9781324035602 (cloth) |
ISBN 9781324035619 (epub)
Subjects: LCSH: Eugenics—History. | Eugenics—Moral and ethical aspects.
Classification: LCC HQ751 .R87 2023 | DDC 363.9/209—dc23/eng/20220808
LC record available at https://lccn.loc.gov/2022027417

ISBN 978-1-324-06613-2 pbk.

W. W. Norton & Company, Inc., 500 Fifth Avenue, New York, N.Y. 10110
www.wwnorton.com

W. W. Norton & Company Ltd., 15 Carlisle Street, London W1D 3BS

1 2 3 4 5 6 7 8 9 0

Dedicated to my friend Marcus Harben
(March 5, 1974–February 11, 2021)

and his beautiful legacies, Jazmin and Joseph, whose existence
makes this a world worth fighting for

CONTENTS

A NOTE ON TERMINOLOGY

This is a book about the history of a political ideology, and its repercussions in our time. The foundations of the idea lie in our eternal interest in heredity, which today is governed by the science of human genetics. This is a relatively new field, one currently enjoying a golden age as we continue to make headway in understanding the underlying biology of people. It comes with plenty of jargon and terminology. Some terms you will be familiar with, but for our purposes require definition. Alongside the history, I will be talking about genes, genomes, chromosomes, DNA and proteins, and though knowing how these work is the basis of high-school biology, people like me spend a lifetime discovering that ours is a science of exceptions, and the rules of biology are always stretched, and sometimes broken.

Scientists, argumentative as we often are, frequently fail to settle on definitions that satisfy one and all. President Harry Truman is supposed to have said, "Give me a one-handed economist!" so that his adviser could not offer an opinion and follow it up with "on the other hand. . . ." The same plea could be leveled at my ilk. Nevertheless, here goes: DNA is the name for the molecule that carries genetic information, and typically is depicted in the iconic double helix. Genes are pieces of DNA that code for proteins. Proteins are the workforce of living things: all life is built of or by proteins. Genes are part of chromosomes, which are long stretches of DNA harboring many, sometimes thousands, of genes. Chromosomes are also built from lots of DNA that control the regulation of genes, that is, switches that say when and where they need to be active. All organisms have a set number of chromosomes, and in humans that number is typically forty-six—twenty-three of these come from each parent, and twenty-two of these are paired, containing different versions of the same genes. The remaining chromosomes are the X and Y: women typically have two X chromosomes, men have an X and a Y. The genome is the total amount of DNA in an individual, or species, which includes all the genes, all the control switches, and more, much of which we don't really yet understand. The genotype is the particular versions of genes an individual has; the phenotype is how it manifests in the physical body.

These are the basics of human biology, and I shall try to limit the technical language. However, eugenics was a political ideology that was shackled to genetics, and in some cases there is no way round grasping that nettle.

And while these basics are taught as high-school biology, there are infinite levels of qualification, exceptions, caveats and finicky details that are far from trivial. This is why human genetics is not finished, nor will it ever be, and why those who confidently assert truths in science are often buoyed by ignorance rather than knowledge.

CONTROL

INTRODUCTION

I f you have children, or are ever planning to, you will surely
want them to live well. You hope that they are free from disease, and that they fulfill their potential—in school, in physical fitness, in their allotted share of happiness—and that they
will not suffer pain. What are you willing to do to ensure this?

On November 28, 2018, I awoke to find my phone battered
with thirty-three new text and voice messages. It was mercifully
only the second time this had happened, the first being when
my telephone and email had been published on an American
White supremacist website, with the unsolicited invitation to
"contact him." This happens when you write about race.

This time was different though and, perversely, more troubling. The messages were all requests from a harried media,

asking for my comments on something that had happened in Hong Kong while I was sleeping. At a medical conference, a scientist had announced the birth of twin baby girls that he had genetically modified as embryos and reimplanted into their mother. As far as we know, Lulu and Nana (their given pseudonyms) were the first gene-edited humans ever born.

My colleagues and I in the scientific and media worlds struggled to get our heads round just what had happened. Where were these babies, and were they healthy? How had this scientist bypassed the international laws and agreements that expressly forbid such practices? There was a scramble to verify the scant information that the researcher himself had released to the public, via a presentation at the conference and a couple of slick YouTube videos. Professor He Jiankui, a biophysicist from the Southern University of Science and Technology in Shenzhen, had attempted—and by his own data, failed—to introduce a naturally occurring genetic mutation into two fertilized human embryos. His intention was to grant immunity to HIV infection to the children who would grow from those clumps of cells. The technique he had used is called CRISPR-Cas9, a gene editing tool that over the ten years since its invention has radically changed our ability to control the source code of biology. The father of the girls has AIDS, Professor He revealed, and the proposed genetic rewrites would theoretically protect the man's children from the risk of HIV infection for life. A third child—we now know she is called Amy—was also confirmed to have been born, though further details are even more opaque.

He Jiankui was widely and immediately condemned for this human experimentation. It *was* experimentation, not therapy or medical treatment or intervention. He wasn't trying to cure a disease or treat a condition. Each of the girls' genomes was indeed modified in this procedure, but neither change resulted in the genetic variant that naturally bestows the bearer with HIV immunity. The proposed edits had failed, and he had effectively conducted a genetic experiment on those children, whose fate and health is still not publicly known.

A frenzied press called him a "mad genius" and "China's Frankenstein." The term "genius" is inappropriate here; the attraction of the CRISPR-Cas9 tool is that it is designed to be easy to use—a genetic edit in a lab animal that might have taken years to achieve two decades ago can now be done in a few days by a student, with much greater control. The embryos were examined, sampled, selected and reimplanted with techniques used in reproductive clinics and genetic diagnosis labs thousands of times a day all around the world. He had merely combined two fairly standard techniques in genetics and reproductive medicine to do something that is morally and ethically reprehensible, and indeed criminal.

As for "China's Frankenstein," knowledgeable readers like to point out that Mary Shelley's Frankenstein was the scientist and not his unnamed creation. But wise ones know that Dr. Frankenstein was the monster. The ethics of experiments on humans are well established, and necessarily constrained. Experimentation of this sort is broadly prohibited under various conventions, including the Nuremberg Code and the Declaration of

Helsinki. The story of what happened next is shrouded in mystery, but we know that a year later He Jiankui was sentenced to just three years in jail and a fine of three million yuan (around $470,000). He was released in March 2022.

The legal, ethical, moral and scientific issues thrown into relief by the birth of Lulu and Nana are enormously important, and later in the book I will examine this tragic and wicked story in detail. The principles in play are not new, but the technology employed by He Jiankui to put them into practice is. Questions about modifying life, and human life specifically, are at least as old as *Frankenstein*—Mary Shelley's novel was inspired by the then new sciences of physiology and galvanism. Throughout the twentieth century, fictions and fears grew alongside the field of genetics as it blossomed and revealed the underlying software in which life is written. In the postwar era, technological advances such as the contraceptive pill helped charge a social revolution where sex could be decoupled from conception, and women could claw back some measure of reproductive autonomy. The invention of in vitro fertilization in the 1970s has resulted in the birth of several million people who previously could never have existed.

Science continually tinkers at the edges of the known, unpicking the fabric of reality, sometimes with a specific purpose in mind, more often fueled by slakeless curiosity. We invent tools to help people and answer questions about the universe, and everything in it. But science does not operate in a vacuum, and its purpose is ultimately in the service of humankind.

We are technological creatures, and we have been continuously crafting and designing nature to serve our needs and desires since long before our species emerged on the landmass that we now call Africa. Genetics is a new science, a century old at most, and really just a few decades old in any meaningful sense. The fusion of genetics with evolution is similarly only a twentieth-century field of research, as we came to understand genes as the mechanism by which evolutionary change occurs. But crudely, genetics and evolution are only the study of sex and families, which have been the primary fixations of humans since before the origin of our species. People like me examine sex lives and families at a resolution that removes every drop of joy inherent in both: at a molecular level, across populations and over oceans of time. In doing so we have made great leaps forward in understanding how human heredity works.

However, all science is political. This is a statement that causes vexation among some who confuse the ideals of science with its reality. We aim for an objective description of the world, and try to minimize the grubby political, personal and psychological biases that hinder our view of reality. But in all science—and especially the scientific study of humans—we inherit knowledge infected by the contingencies and political obsessions of our scientific forebears, whether we know it, deny it or acknowledge it. Sometimes the biological and the political are deeply intertwined. For example, in the Neolithic, we sought to control livestock and plants in order to ensure a regular supply of food, thus inventing agriculture and farming, and with that trade and commerce, and the foundations of civilization. In doing so, some pastoralist societies inadvertently selected a genetically encoded ability to digest milk after weaning, which gave them a rich source of food that had been unavailable to their lactose-intolerant ancestors. Some seven thousand years later, this dairy schism became a key distinction between the Hutu and the Tutsi, imposed by German and Belgian colonists seeking to sow racial disharmony. Indeed, some White suprem-acists today mistakenly think this evolutionary adaptation to dairy farming is a mark of European purity, ignorant of the fact that it also arose in Kazakh, Ethiopian, Khoisan and Middle Eastern pastoralists who also evolved alongside the mammals they milked.

Biology is political in a recent historical sense too, tethered as it is to the seventeenth-century invention of race as a means of human classification and—the subject of this book—the sci-entific attempts to control human biology and, with it, society. Our fundamental biology is not isolated from the architecture of our societies, nor has it ever been. We are bound by the trap-pings of our fleshy hardware, and evolved to push against them. This is the human condition. Humans are a paradox of nature, freed from the shackles of natural selection by our big brains and big societies, but still wedded to the mechanics of sex, inheritance and genetics.

In principle, we are free to choose with whom we mate, and our biological imperative to do so is diminished and controlled, at least compared with all other animals. We can decide whether to have children or not, how many, and how we raise them. We have options on deciding whether to proceed with a particular pregnancy when there is a known risk of disease or suffering in that future child, at least in many parts of the world. In the age of gene editing, we can even tinker with their DNA.

This is a book about two forces that shape us: control and freedom. These are such potent ideas that they serve as tectonic plates on which civilizations are built, but biology is frequently missing from our discussions about our lives and those of our children. It cannot be ignored. Life and liberty are two-thirds of the unalienable rights decreed in one of the greatest manifestos ever written—the United States Declaration of Independence. The hallowed principles in its preamble are so fundamental that they are described as self-evident truths—people are created equal, as authorized by a God, and life, liberty and the pursuit of happiness are enshrined as natural law. They are therefore incontestable, and the inarguable basis of governance.

They are of course fictions, noble lies. Set aside the fact that the men who wrote these words owned other humans as possessions, to be sold and to live their lives enslaved, much of which was justified by a newly invented science of biological classification. Slavery is a red stain in humankind's ledger and its legacies persist: many people around the world remain enslaved today. Nevertheless, the ideas spelled out in the Declaration of Independence are beacons of light, fundamental entitlements afforded to humankind. They are also ideals that we have never achieved.

People are not born equal. They are conceived already cuffed to forces beyond their control that will shape their lives, limit their opportunities and keep their ability to fulfill those unalienable rights beyond their grasp. Class, race,

wealth, nationhood, biology and randomness are all con-
founders to the principles of equality. You were not born free
of these forces.

Biology and society are inseparable. Our biology is *enacted* in
society. This is an obvious thing to say, but I think it's often over-
looked. Society emerges from our biology, and from the interactions between these evolved bodies that we inhabit. We often
deploy the clumsy ideas of nature and nurture to describe what
is innate in us, and what is extrinsic. What this really means is:
genetics (that is, what is encoded in DNA), and everything else
in the universe. Your genome is a script, etched into the kernel at
the center of your cells, but the film of your life is played out in
the countless forces that determine how that script is performed.
Nature was never *versus* nurture; it is and always was *via*.

For the whole of history, all cultures, all countries, all societies, have considered the principles of who can reproduce, who
lives and who dies. Governments, society, biology, tradition
and myriad other factors nudge and steer and compel people
away from the freedom to reproduce with whomever they want.
Biology and culture are inextricably entwined: each sculpts the
other. For just over a century, we have referred to the deliberate
crafting of society specifically by biological design with a word
that for half of its existence has been regarded as desirable, and
for the other half, poisonous: *eugenics*.

Eugenics is a project with a short history, but a long past. The
oldest readers will have direct memories of the Second World
War, and how governments tried to exert the most pernicious
forces of control on their populations. Eugenics is perhaps most

closely associated with the deranged acts of the Nazis and their evil attempts to exterminate not only millions of Jews, but also hundreds of thousands of people with physical disabilities or mental illnesses, or other characteristics such as homosexuality. They were collectively categorized as *Lebensunwertes Leben*—"lives unworthy of life." The escalation of Nazi Germany's eugenics program to the Final Solution occurred in incremental steps, preceded by years of broader policies aimed at the general improvement of the German people under the guise of *Rassenhygiene*—"racial hygiene." The toxicity of the idea of eugenics no doubt emerged from our collective discovery of the horrors of the Second World War, but state-sanctioned eugenics policies were also implemented in more than thirty countries, and some of these endure in the twenty-first century. They were espoused enthusiastically by the two great opposing powers of the postwar twentieth century—the communist Soviet Union and the capitalist United States. Eugenics has always enjoyed bipartisan support.

Eugenics is in many ways a defining idea of the twentieth century. It was enacted as policy by the most powerful and populous countries on Earth and fueled tyrannical regimes that tore the world apart with unprecedented vigor. Before that though, eugenics was a guiding light for the betterment of Western societies, viewed as normal and desirable by people across political divides, and forcefully championed by the most powerful men and women in society. Winston Churchill was a key driver of eugenics policy in the United Kingdom in the first two decades of the twentieth century, as was Theodore Roosevelt in the

United States. Margaret Sanger, a pioneer of reproductive rights for women, advocated for eugenics policies, as did the scholar W. E. B. DuBois, as a potential mechanism for racial uplift for Black Americans.

Many playwrights, suffragists, philanthropists and philosophers, as well as more than a dozen Nobel Prize winners, embraced the ideas of eugenics as a force to change society, some with an almost religious fervor. The first part of this book is a history of an idea that hid in plain sight, from its roots in key philosophical texts of the classical world, in obscure and popular scientific books, and all the way into its genocidal realization in the twentieth century.

It's difficult for us to comprehend, only a hundred years later, quite how ubiquitous this idea was in the early decades of the twentieth century. But the evidence is right in front of us, baked into our culture and literature. Novels such as Aldous Huxley's *Brave New World* or H. G. Wells's *The Island of Dr. Moreau* are tales about genetic manipulation of human life, and the scientist and eugenicist Julian Huxley (brother of Aldous, friend of Wells) even advised on the 1932 film adaptation of *Moreau*, entitled *Island of Lost Souls,* for its scientific accuracy. Eugenics percolated through culture in less obvious and overtly fantastical ways too. A seam running through *The Great Gatsby* is the then popular pseudoscientific fear about the replacement of ruling classes by less desirable members of American society—immigrants, African Americans, Irish, the poor—an idea that fueled much of the development of eugenics policies in the West, and persists among White supremacists to this day. "The Jews will not replace us!" screeched hysterical Nazis in Charlottesville, Virginia, in 2017, in full view of the world's media, though it was never clear whether they imagined that they were to be replaced by Jews, or that Jews were orchestrating their replacement. Either way, the long-standing fantasy of the threat of population replacement is part of the elusive promise of eugenics—to exert control on who lives, who dies and whose people should be preserved.

The cultural ubiquity of eugenics even extends to our food: John Harvey Kellogg was the effective creator of the cornflake, and with those cereals reinvented breakfast for large parts of the world. Many readers will have eaten a dried cereal crop this morning (with a splash of milk enabled by your lactase persistence mutation) whose evolution began with Kellogg's weird, obsessive desire to control libido with bland foods, and thus protect and preserve the precious bodily fluids of upstanding American men. Ridiculous though this sounds, Kellogg used his immense wealth to support and develop his obsessions. He became a champion of eugenics in the United States and, with that, a key driver of principles of racial hygiene that grew in parallel with the same policies in Nazi Germany. Until the Second World War, eugenics was a beacon of light for many countries striving to be better, healthier and stronger.

The second part of this book concerns where we are now. Population control policies that resemble eugenics are still enacted today. Toxic though the word may now be, the ideas underlying eugenics are not historical. They are our present. We live in a time where forced sterilization of women is ongoing in many nations, and where sex-selective abortion is rife in the two most populous countries on Earth, India and China. The Chinese one-child rule implemented in 1979 evolved into a two-child policy only in 2015, and three in 2021. But in 2010, it was modified with the Iron Fist Campaign—the compulsory sterilization over the course of three months of ten thousand women who had violated the law by having more than one baby.

An American campaigner against drug addiction buys the fertility of addicts: Barbara Harris has used the charity she founded, Project Prevention, to pay three hundred dollars each to American drug addicts and alcoholics to have long-term contraception or sterilization to prevent their having children born into substance abuse. According to the group's own statistics, she has paid 7,600 people for control over their reproductive biology.

Today in Saskatchewan, Canada, there is an ongoing class action in response to the coerced sterilization of hundreds of First Nations women, as recently as 2018. And in the United States, an estimated twenty women underwent involuntary sterilization in Immigration and Customs Enforcement detention centers in 2020. These are just a handful of examples from dozens, but they clearly demonstrate that even in our enlightened world, reproductive control is frequently exerted unilaterally from state to women and men, and our cherished freedoms become specters.

In contrast to these atrocities, we now have the ability to examine fertilized embryos and test them for the existence of genetic disorders such as Huntington's disease or cystic fibrosis, and then implant only ones that are free from these curses. This technique, coupled with gene editing, was what was used by Professor He Jiankui in his misguided attempts to engineer babies immune to HIV infection.

We routinely screen pregnancies for conditions such as Down syndrome, and offer women the choice of terminating that pregnancy. In Denmark, access to early screening for Down syndrome is available to all women regardless of age, and around 95 percent of women opt for an abortion if the syndrome is detected in the unborn fetus. In 2019, only 18 people with Down syndrome were born in a population of 5.8 million, compared with around 6,000 in the United States.

Are these techniques eugenics? I don't believe that they are, though both eugenics and these reproductive technologies share history and scientific ancestry. These are medical interventions designed to offer options to parents, enabling them to make choices about the medical health of lives they may want to bring into the world, whereas eugenics in its original form was designed to sculpt societies through selective breeding. When it comes to prenatal screening and embryo selection, the question of whether this course of action is the right one to take remains. Nevertheless, we will see in these pages that the techniques invented to treat genetic conditions evolved from the very laboratories—including the one in which I was an undergraduate and am currently a lecturer—that first nurtured the ideas of eugenics.

The language we use in these conversations frequently is confusing, lacks clarity or fuels political battles. Eugenics is such an irredeemably toxic idea today that it gets thrown around as an insult to scientists who do any work in the field of human or behavioral genetics, especially where it relates to social status, ancestry or cognitive abilities. The word "eugenics" itself is a neologism, invented by the Victorian polymath Francis Galton, who went on to define it in various ways, all to do with the molding of populations according to the desirability of particular characteristics: it is a fusion of the Greek prefix *eu-*, meaning good, and *genos*, meaning offspring—well born.

The formal origin of eugenics is inextricably bound to the birth of genetics, and the study of inheritance. Galton founded the Eugenics Records Office in London in 1904, which institutionalized this movement, and tied it to University College London (UCL) for many decades to come. The men who followed in his footsteps at UCL were similarly brilliant thinkers on whose shoulders entire disciplines rest. Some of them also held deeply racist views and believed fervently in the principles of eugenics. Their opinions are often grotesque in our age but were not atypical for theirs. Today, their legacies collide.

In an era in which we confront the public representations of historical figures, we are beginning to reassess these men of science. Galton and his primary disciple, Karl Pearson, both had their names removed from buildings in June 2020 following the first official inquiry into UCL's eugenics past. I know this history because I studied as an undergraduate in the Galton Laboratory at UCL and was taught by the Galton Professor in the Galton Lecture Theatre. I, along with dozens of others, am only three direct academic generations removed from Galton himself. Since the age of eighteen, my work and interests have been profoundly influenced by the colossal scientific legacy of Ronald Fisher, a man who invented countless statistical techniques that are used in many disciplines and formed the basis of modern evolutionary biology. Fisher was also a committed eugenicist throughout his life and maintained ties with a Nazi scientist who had worked alongside Josef Mengele on human remains from Jews murdered in death camps.

Though UCL's association with eugenics is unique, it is by no means the only nucleus where this ideology bloomed. The salons and clubs of the wealthy and privileged around the Western world were hubs for much of the discourse about how to improve a people, and how to use this new science to cling to power. But much of this chattering class conversation was fueled by academic institutions, and their impact on men of influence. The Ivy League and other top American universities all have historical associations with eugenics. Princeton played host to several key advocates, including Harry H. Laughlin, who successfully campaigned for eugenics in America and Nazi Germany (from whom we will be hearing more later); Carl Brigham, who advocated for the intellectual superiority of the Nordic race and later designed the SAT; and the president of the university from 1902 to 1910, Woodrow Wilson. In 1911, one year before winning the presidency of the United States, while serving as the governor of New Jersey, Wilson signed legislation to "act to authorize and provide for the sterilization of feeble-minded (including idiots, imbeciles and morons), epileptics, rapists and certain criminals and other defectives."

David Starr Jordan was the president of Indiana University in Bloomington and went on to be the founding president of Stanford. He was also a long-standing eugenicist who believed war would strip nations of the best and leave only the survival of the unfittest. Jordan served on the board of the Human Betterment Foundation, a think tank comprising academics from UC Berkeley, Harvard, Caltech and the University of Southern California that advocated for compulsory sterilization legislation in the United States. Jordan's nominal legacy has recently been expunged from the campuses at both Stanford and Indiana.

I'm not in the business of erasing or ignoring knowledge because of its provenance. We rightly celebrate the pleasing Newtonian principle that we see further by standing on the shoulders of giants. We are not nearly as good at recognizing that our vantage point can be unstable because those giants may also have been bastards. This is how history works. It is not there to reassure us or make us feel warm. If the actions of our forebears only make you feel proud, and don't sometimes baffle, upset or anger you, then you may not be doing history at all.

There was and is, in practice and theory, positive and negative eugenics, where desirable traits would be encouraged and undesirable traits purged. The control exerted on populations via these means predates our age by millennia, as I'll discuss in the first part of the book. Recognizing that it is an ancient and arguably ubiquitous practice does not excuse it, nor exonerate those who enacted policies of selective breeding in humans or murdered people deemed unworthy. Nevertheless, there is real discussion today about the genetic enhancement of traits in children and populations that are indubitably descendants of the eugenics policies of the Edwardian era.

In the last few years there has been an exponential increase in understanding of the human genome, as well as DNA's role in sculpting our lives, across physical, psychological and cognitive traits. When you toss that into the mix with novel tools in precision gene editing, and reproductive techniques such as embryo selection, what emerges is an urgent necessity to make discussions of eugenics part of our public discourse. Today, with these new techniques becoming more readily available in labs and clinics, a new conversation about ranking people according to intellectual abilities and about the possibility of selecting embryos for desirable traits is emerging, and often the tenor of this discussion appears oblivious of a history that is not much more than a century old. Eugenics has a complex history and features incredibly complex science. Yet it is a conversation that we must have, and to discuss it adequately, we must be armed with facts, knowledge, a long view of history and the voices of those who would have been marginalized or worse by the policies of the past.

We must also be intellectually honest. These are inflammatory ideas, and we live in a febrile age. I am no eugenics apologist and have no desire to appear controversial by trying to reclaim this word from its toxic history. The purpose of this book is to clarify and expose.

It's a deliberately short book. The full story of eugenics would incorporate an almost complete history of the twentieth century as well as thousands of years of attempted population control that preceded it, including in the dozens of countries and cultures that enforced policies such as sterilization, infanticide and genocide. Instead, my intention here is to carve a particular pathway through history in pursuit of this idea, and to examine it honestly. That means facing up to some of the worst atrocities committed by humans on other humans, which means that this book is not neutral. A "balance book" view of history is not a smart way to understand the past, and many of the players—scientists, writers, artists, politicians and presidents—in this story are neither goodies nor baddies, but complex people with a range of views.

Some readers might feel discomforted by books primarily about science, written by a scientist, that contain both political analysis and judgment. However, human sciences are not outside of politics, no matter how noble an aim that might be. A book about race, genetics or eugenics cannot be politically neutral.

At this point in time, when conversations about eugenics are returning to public life, a sound understanding of the underlying science is urgently needed. The actions of the Nazis were evil, and the ideas of eugenics in the past are now morally repugnant. But we should also consider a seemingly simple question: Would eugenics *work*? The stated aims of the eugenicists were to remove undesirable qualities from the populace and encourage desirable ones. With a modern understanding of genetics, and with the enacted legislation of various countries, we should be able to determine the outcome of eugenics policies, and therefore determine how effective the strategies of population control were, or could be. It's not a straightforward question to answer though. The superficial and often ignorant public discussion will tell you that obviously it would—surely a geneticist such as myself could not deny that humans are evolved, and our biology is encoded in DNA, which can be selected artificially as it has been naturally since life began. We have loosened the shackles of nature, but we are not free of them, and we are not immutable beings.

That is obviously true. But the question of the success of eugenics programs is much more complex, subtle and impenetrable than a bluff "of course." We are not farm animals, our genomes are poorly understood and no other creature is as dependent on the nongenetic world that our DNA sits within, which makes the question more difficult to answer, far from settled and, from a purely scientific point of view, much more interesting. America, Germany and many other countries were confident enough in the science of heredity that they based policy on it: but did it *work*? Could it have worked if allowed to grow? Recall Charles Darwin's phrase that cannot be repeated often enough: ignorance breeds confidence more often than does knowledge.

What follows concerns us all, because we are all part of humankind. I have focused on the development of eugenics in three countries, Great Britain, the United States and Germany, though it happened—and continues to happen—all over the world. And I approach this subject as a scientist, one attempting to understand the role of science in a complex global history. Families, sex, society and our knowledge of biology—and how we put that knowledge to use—have and always will shape us as a species. Biological control has been exerted by the powerful for thousands of years, on populations, on women by men, on the powerless and on the lives of people deemed undesirable, defective or simply enemies. The discussions in the age of the genome are recapitulations of the eugenics of a century ago, whose enactment dominated the twentieth century. To know this history is to inoculate ourselves against its being repeated.

PART ONE

QUALITY CONTROL

This is the story of an idea that we don't really talk about. Even so, it has been part of our societies for millennia, and is fundamental to the story of humankind. It begins with Plato.

Republic is one of the founding texts of Western civilization. Plato's dialogue with Socrates (and other Athenians) considers the structures of society, of justice, the nature of the soul, happiness and many other themes of the world of humans. In books V and VI of *Republic*, Plato outlines a detailed plan to control the breeding of the people in a utopian city-state, where a guardian class would forgo their wealth, and women and men would be matched by the state according to their qualities, like sporting dogs or horses bred for their strength or speed. Gold-standard women would be paired with gold-standard men, or bronze with bronze, as assessed with a numerical quantification. Inferior children would be relegated to a working class and discouraged from breeding. Plato was concerned with declining birth rates, so suggested that the population size would be maintained at around five thousand by the imposition of marriage festivals, as decreed by the ruling philosopher-king. Children born with defects would be hidden away, which may well have been a euphemism for killed. *Republic* was written in the fourth century BCE, but Plato's plan for the ultimate city-state never came close to being enacted.

Perhaps therefore, this story really begins not with Athenians, but with their sworn enemies in the Peloponnesian Wars, the Spartans. That warrior people, famous for their epic if ultimately futile stand against the Persians invading from the east, valued militaristic strength above all. Boys began their training aged seven, but only if they had survived the first selection hurdle, which occurred at birth. Newborn babies were tested for physical fortitude by dunking them in wine, and any showing weakness or deformities were thrown off Mt. Taygetus into a chasm known simply as Apothetae—"the Deposit."

Or maybe that tale is a myth. It's reported only by the historian Plutarch, centuries after the fall of the Spartans, who remain lauded as a model of a society defined by might, despite ending their legendary era literally as circus performers. There is no physical evidence in that pit today, no bones of babies, only those of adults. Perhaps they were carried away by animals—or maybe this child winnowing never happened.

In the Roman ruins in Hambleden, Buckinghamshire, in the United Kingdom, the bones of ninety-seven babies and children were uncovered in 1921, which some historians have interpreted as evidence of a version of Spartan infanticide. But it may just reflect differences in burial practices—adults cremated, children interred. In Rome though, at the zenith of its power, the matter was uncontroversial. The Stoic philosopher Seneca was quite clear on the state policy of infanticide, writing in the first century CE: "We put down mad dogs; we kill the wild, untamed ox; we use the knife on sick sheep to stop their infecting the flock; we destroy abnormal offspring at birth; children, too, if they are born weak or deformed, we drown. Yet this is not the work of anger, but of reason—to separate the sound from the worthless."

Or do we instead begin in Iceland, in the year 1000 CE, at the point when the pagan Norse were flipping their religious affiliation from Odin, Thor and Hella to Jesus Christ and the god of the Bible? The *Íslendingabók* is the key text that documents the first few centuries of Iceland's story, including the pivotal moment when one chieftain, Thorgeir Ljosvetningagodi, successfully argues that island-wide peace can be achieved only if Icelanders accept the blood of Christ. A concession—one of only three—offered to the Vikings in exchange for their baptism was that the old Norse laws allowing exposure, and therefore death, of unwanted, often deformed, newborn children would remain in force under their adopted Christianity.

Anthropologists in the twentieth century reported on various infanticide practices from many cultures. In Africa, some hunter-gatherer populations, such as the !Kung San from Botswana and surrounding areas, were thought to have killed their babies when resources were scarce. Some Inuit people were believed to perform female-specific infanticide during breast-feeding to increase the chances of the mother's becoming pregnant with a son instead. Nineteenth- and twentieth-century researchers reported that the killing of deformed or premature babies was common among Aboriginal Australians—as well as the infanticide of twins, who, according to the anthropologists studying these peoples, were believed to have been fathered by two different men. Though outlawed in both countries, in India and China, the killing of baby girls persists to this day as a means of ensuring economic and social status within families.

However uncomfortable it is to acknowledge this, infanticide has been a constant feature in human societies. Our modern views enshrine the sanctity of human life, and a desire to preserve it however we can. We seek to reduce suffering and enable the lives of people who in the past have been marginalized or, worse, not permitted to live.

Much of what follows is horrific, and some is absurd. The language we use to describe eugenics is imprecise and comes laden with inherent value judgment: we use expressions like "population control," "birth control," "eugenics," "social engineering" and "social Darwinism," none of which is politically neutral. As you read that last sentence, did you think nothing from an emotional or political perspective? Such words and ideas also sit uncomfortably among other movements that carry with them a history of violence and are even more emotionally electrified:

racism, fascism, genocide, empire, White supremacy and other ideas that are serious politics.

I'm a scientist, and we are a tribe whose pursuits, in theory, serve a higher purpose. We constantly strive for real truth, via our tried and tested methods that seek to supress the political, or the subjective, and amplify a reality that exists independently from these flawed minds and feeble bodies that we inhabit. What we do, in principle, transcends politics and morality.

Scientists can tell themselves this lie as often as they like, but it will never be true. When we talk about the control of lives, the question of who gets to live or reproduce, we are in a territory where biology and politics are inseparable. That term, "eugenics," was coined in the nineteenth century, and it was designed to mean "good birth" or "well born"—a statement of intent that our objective should always be to encourage the health of individuals in the next generation. Are we not all eugenicists in the purest sense? No one wishes a child to be born into pain, or for a people to be beset by suffering.

The trouble is that this idea has always been—and perhaps can never not be—locked in step with deeply illiberal positions, where freedom is curtailed, choice expunged and control imposed by the powerful on the weak. Racism, sexism and classism are inherently built into attempts to make people fitter, happier, more productive.

History asks us not to judge the people of the past by our standards but by theirs, in order that we may understand the culture and context in which they operated. Some of the language and phrases of the eugenics movement remain in use today but have evolved far away from their original meanings: today's casual insults such as "imbecile," "moron" or "idiot" carried specific psychiatric significance a century ago, and were part of a lexicon used in the diagnosis of mental characteristics that could warrant enforced institutionalization and, in hundreds of thousands of cases, involuntary sterilization.

Similarly, many words from this history can spur confusion because of their own evolution. The most obvious example is "race." Today, we regard race as an important categorization that is determined primarily by social conventions, but it has no meaningful or useful underlying biological basis. This modern understanding of human variation based on the science of population genetics is uncontroversial within the various fields of human biology. The extensive use of the word "race" in the nineteenth and early twentieth century bears little relationship to this modern understanding. Race was generally used to mean "type," or sometimes "population" or "kin"—Darwin talks at length about races of cabbage and pigeons in *On the Origin of Species*. This usage may have some limited overlap with how we talk about race today but it is not the same.

Eugenics was not fundamentally a racist ideology, except in the countries where it was by default—notably, the United States and Canada. What this means is tricky to unpick, but the basic ideology of enriching a population via selective breeding becomes racist only when it is fused with the assertion of White superiority. In Britain, this idea was welded to colonial expansion and empire building, but also was directed toward a growing underclass of White Britons. In the United States, eugenic purification was almost exclusively envisaged for the ruling classes of White European–descended Americans, and it involved the suppression of traits deemed undesirable in others. But in a society stratified much more unambiguously along racist lines, this effectively meant that eugenics policies disproportionately affected racialized groups, primarily those descended from the enslaved and Indigenous Americans. So while the basic principles of eugenics did not necessarily have racism explicitly stated within them, in America, it was impossible for eugenics to be anything but an expressly racist ideology.

In the United States, popular books assumed White superiority with the adoption of the concept of Nordic purity: that northern Europeans were a particular race who were simply better than others, and Black people and Indigenous Americans were inferior and from immutable inferior stock. The Nazis took this idea to absurd lengths with their extreme genocidal racism. The outcome therefore of most eugenics policies was racist in particular ways, and racialized groups were—and are—disproportionately affected. Institutions such as John Harvey Kellogg's Race Betterment Foundation were not necessarily focused on race as we see it today, and the most extensive implementation of the ideas of "racial hygiene" by the Nazis was not limited to Jews, or Roma, or groups that we might now call races. Like many of the supporters of eugenics in the nineteenth and early twentieth century, the Nazis' aims were to ordain the purification of their own ill-defined group against all others.

With eugenics, we stand at the interface of biology, morality, feminism, racism, politics and an ever-changing global stage.

The theories and methods that people have deployed in order to exert control on who gets to reproduce, and who is reproduced, are wide. Some are murderous and genocidal; many infringe what we today would consider basic principles of human rights; others are preposterous—schemes and grifts invented by political con men whose moral compasses in any age are well off kilter. All concern sex.

We, like most animals most of the time, are sexual beings. There are a few exceptions in nature, but for the most part, that means that two individuals with different genetic and biological characteristics—a female and a male—from one species are required to produce offspring. The forces of nature that determine who copulates with whom are well understood for animals. If we take the very human trait of free will out of the equation, and dial up the biological imperatives of mate choice, the mechanics of sexual selection reveal themselves. The world of sex is directed by females producing a few large immobile eggs, and males generating a multitude of tiny but motile sperm. Couple that with a spectrum of parental care, mostly (though not without exception) an investment paid in majority by females over males. With these principles, many of which were described by Charles Darwin, and developed by the evolutionary biologist Ronald Fisher—from whom we will be hearing much more in the forthcoming pages—one of the cornerstones of evolution is planted in the ground.

They explain why males and females look and behave differently. They explain why males fight to seek the approval of females, who will subsequently choose the male that will sire their offspring. They explain why males often have ludicrous plumages and ornaments, or engage in preposterous shows of nothing but bravado and swank, again to impress a choosy female, who has evolved to adjudge that these costly displays may showcase genes better equipped to endure in their children. In recent years, we have discovered that males of some species deploy all sorts of mechanisms to prevent females from mating with other males, in an attempt to guarantee that they become the fathers over their lesser brethren. Sometimes this is crudely physical—a dog's penis will expand after ejaculating and lock him inside a female for as long as possible. Other techniques include releasing sperm that form plugs, or ejaculate that is semen-rich as a means of flushing out a previous male's sperm. Females have their own tricks too—many ducks have labyrinthine vaginas, tortuous with blind alleys. Some female butterflies have an extra stomach in their vaginas—the bursa copulatrix—that has evolved to digest sperm from lesser males. Sex has evolved in animals to serve the central principle of evolution—that is, the replication and endurance of genes into the next generations, each sex exerting its own attempts at controlling the outcome of a highly competitive field.

We are evolved, there are two sexes, and the mechanics of copulation, eggs and sperm, fertilisation, pregnancy and birth are not wildly different from our animal cousins. But of all the fascinating tricks that other animals deploy to extract control over their reproduction, we do none, and our mechanisms of biological control are very different. Eggs are the biggest cells in the human body, and sperm the smallest. A female is born with all of her eggs already in place, formed as they were while she slumbered in the womb of her mother. They sit, frozen in stasis until the onset of puberty, and each is activated, one by one, until the onset of the menopause, after which no more of the eggs are released. Eggs therefore are precious: only a clutch of them may become a child, and most are discarded during each one of a limited number of menstrual cycles.

Sperm have no such transgenerational splendor. The single sperm that made you matured in your father's testes within a couple of days before it met that egg, alongside billions of others—trillions during a lifetime—all of whose existence was cheap, short and uneventful.

That part of biology is pretty much the same in all animals, but in the last few tens of thousands of years, we have radically dialed down the imperatives of evolution, and replaced them with a semblance of choice. Most people are broadly heterosexual in thought, word or deed, and typically pair with a member of the opposite sex, with whom they have children. But as you read that sentence, I will wager that you may well have been considering the many people who don't fit that pattern. Nonreproductive sexual behavior is common throughout nature, but there are vanishingly few examples of animals that choose not to reproduce altogether. In humans there are millions who elect not to conform to the biological norms of nonhuman animals: people who exclusively have sex with members of the same sex, people who abstain from sex altogether, people who choose not to have children, or adopt, or foster, or any number of variations on how we choose to live our lives that don't tally with the sex lives of other animals. If those demographic categories are proportionally small compared with the heterosexual majority, consider that there are a lot of us, and a small proportion can be a very large number. Accurate accounting for sexual preferences is not easy: numbers vary through time and according to different definitions, but a broadly accepted consensus is that around 10 percent of people globally are homosexual, which would mean a constituency of more than 700 million worldwide—more than twice the population of America. And within that populace, some will adopt, some women will have babies via surrogacy, many will change their sexual orientation during their lives.

On top of the rainbow of behaviors we see today, again though heterosexual pair bonding (to give it its romantic description) has been the most common backbone of social structure for large parts of the world for long periods of time, it is by no means the exclusive history of procreation in humans. Polygamy, where one person maintains sexual relations with many, has persistently existed through human history, mostly in the form of polygyny, where one man has multiple wives, and more rarely as polyandry, which is the reverse.

This brief sex education interlude is relevant for several reasons. Sex, genetics, heredity and evolution are all very closely related aspects of biology that determine the outcome of the lives of individuals and of populations. These processes were not understood when Plato outlined his utopian eugenics program in *Republic*, or when Rome and other civilizations enacted theirs. The first human sperm cells to be closely observed were released only in the seventeenth century, when microscopy became good enough, and the first egg in the nineteenth,* and even then the mechanics of what we now teach to primary-school children were largely unknown.

Only in the nineteenth century did we begin to understand how sex and heredity really works, and only in the nineteenth and twentieth did we add evolution and genetics to that bank of knowledge. Though eugenics protocols have been around throughout history, in theory or in reality, it was only in the modern era that the newly emerging biological sciences could be applied to formalize this ancient practice.

The themes and motivations remained the same. In Europe in the early nineteenth century, new technologies fueled a fundamental change in the production of goods, services and the wealth of nations, and these cultural seismic shifts were themselves propped up on the financial rewards of expanding empires. In Britain, even before the Industrial Revolution had fully taken hold, fears about an expanding working class sat alongside anxieties over resources and the growth of an under-class. The clergyman and economist Thomas Malthus is the best known and perhaps most influential thinker on this matter.

Alongside life and liberty, the pursuit of happiness is the third unalienable right in the American Declaration of Independence. Malthus saw population growth as a metric for that. He examined the notion that population growth would exceed resources, that the poor reproduce faster than the wealthy and

---

* Sperm observed by Antoni van Leeuwenhoek in 1677; eggs by Karl Ernst von Baer in 1827.

that these facts would determine the fate of nations and the overall happiness and success of a state. The trap that he predicted was that a happy population grew exponentially, but that the resources available to them grew only linearly. Therefore, states were bound to a cycle of "perpetual oscillation between happiness and misery." Poor laws in Britain had been in force since the Tudor age, to help the relief of poverty at a parish level, but Malthus thought that these policies encouraged population growth by nurturing a fecund underclass. Malthus is harshly caricatured by Ebenezer Scrooge in *A Christmas Carol*; he won't give money to the poor, as he already pays for the workhouse, and "If they would rather die ... they had better do it, and decrease the surplus population."

Sexual abstinence was Malthus's preferred method for curtailing the exponential growth of the British in the nineteenth century, but he also thought that war, famine and pestilence were necessary for reducing poverty in society. His ideas about populations were controversial right from their inception in 1798, but set the scene for birth and population control that would follow in the coming centuries. For Malthus, as for so many of the thinkers and politicians who would advocate population control over the coming decades and centuries, data about people trumped the people themselves.

THE EVOLUTION OF US

The tectonic plates of biology shifted permanently in the mid-nineteenth century, when Charles Darwin published *On the*

Origin of Species. There he outlined how biological evolution actually occurs. The idea that organisms change over generational time had been around before, but in 1859 Darwin offered the mechanism by which it occurred: evolution by natural selection, descent with modification, survival of the fittest.*

Darwin studiously avoids applying his big idea to us, except for a tantalizing teaser of what is to come: "light will be thrown on the origin of man and his history." But in 1871, in *The Descent of Man*, we are the subjects of Darwin's scrutiny, and humankind's position as a branch on a grander family tree is revealed to the world.

Though Darwin cannot, and should not, be held responsible for the eugenics that would follow, within his idea is a spark that lit the fire. Evolution by natural selection reinforced the correct idea that organisms are mutable and change over time when subjected to external forces that provide advantage or otherwise to their innate characteristics. Variation within a population was a key to this process; subtle (or occasionally monstrous) differences between individuals could act as a fulcrum on which selection could gain purchase. We all know the classic examples: the beaks of finches that are specialized to

* This maxim was not Darwin's but that of the philosopher and biologist Herbert Spencer. I don't like it much because it became tautologous as evolutionary biology developed: fitness is a term used to quantify the ability of an organism to contribute its genes to the next generation. Therefore, it sort of means "survival of the individuals that survive." Nevertheless, Darwin adopted it as an alternative phrase for natural selection, and added it to the fifth edition of *The Origin of Species* in 1869.

deal with particular foods on particular islands; the camouflage of countless animals, from stick insects to the peppered moth, famous for evolving during the Industrial Revolution (and Darwin's lifetime) when birch and other light-barked trees were blackened with soot, and so previously rare dark moths became the most common type.

History rarely pivots on the actions of one person, but Darwin's half cousin Francis Galton (they shared one grandparent) did more for revitalizing the ancient practice of population control than anyone else. In the years around the publication of *The Origin of Species*, many had begun thinking about applying evolutionary theory to humans and the improvement of society. But it was Galton who did the most to rejuvenate the idea. Not only did he coin the word "eugenics,"* but his attempts to formalize it scientifically became the enduring passion of his life, to be followed with a religious zeal.

Galton was the archetype of a gentleman scientist. Born in 1822 into wealth (his peace-loving Quaker ancestors were also gunsmiths), Galton was no longer required to work for a living after his father died in 1844. A truncated stint at medical schools in Birmingham and London was followed by four years

* In a footnote in his 1883 book *Inquiries into Human Faculty and Its Development*: "We greatly want a brief word to express the science of improving stock, which is by no means confined to questions of judicious mating, but which, especially in the case of man, takes cognizance of all influences that tend in however remote a degree to give to the more suitable races or strains of blood a better chance of prevailing speedily over the less suitable than they otherwise would have had. The word *eugenics* would sufficiently express the idea."

at Trinity College, Cambridge, studying the mathematics that would dominate his pursuits throughout the rest of his life. He traveled extensively, including to what is now Namibia, as well as Sudan, the Nile, Damascus and Jordan, and wrote a bestselling handbook called *The Art of Travel* based on his adventures. Galton had a phenomenal mind and left a formidable intellectual legacy. He did pioneering and lasting work in forensic fingerprinting, in synesthesia and in meteorology (he effectively conceived the weather map),* and he invented the dog whistle. He did less enduring and more peculiar work on female beauty, attempting to catalogue and quantify it, and secretly scoring women in public on a three-point scale ("attractive, indifferent or repellent") using a special glove armed with pricks concealed in his pocket. His intention was typical: to apply metrics to something previously intangible. His aim was to show where the most attractive women in the United Kingdom were on a beauty map (it turned out to be London, and Aberdeen at the other end of his scale). In all these pursuits, Galton deployed, and in many cases invented, statistical techniques to crunch numbers for whatever project had taken his fancy.†

Francis Galton's legacy is great, but he will and should be remembered primarily for his role in ushering eugenics into the twentieth century. His work spread the ideas of eugenics to

---

* In fact, it was a recording of the previous day's weather, published in *The Times* on April 1, 1875.

† Basic tools of statistics such as standard deviation, correlation coefficients and regression lines were either invented or developed by Galton to process his ideas and add data-driven evidence.

countries around the world. He directly influenced figureheads of the American eugenics movement and was talismanic to men such as Charles Davenport and Harry Laughlin, to whom we will come in the next few pages. It is in the work of these Americans that the Nazis found scientific, legal and intellectual justification for their own genocidal eugenic ambitions.

More than thirty countries had formal eugenics policies in the twentieth century, but here I will focus on two that did— the United States and Germany—and one that did not. Despite Great Britain's role as the intellectual birthplace of the field, it never enacted a eugenics policy. However, the cultural and political appetite for eugenics in the United Kingdom was robust, and this rests largely on the shoulders of Francis Galton.

GALTON'S GENIUS

Enamored with the work of his cousin, Galton was fundamentally interested in heredity in people, and the potential for their betterment. Darwin's *Origin of Species* concerns evolution, which has no valence toward improvement in any abstract sense, only adaptation. Nevertheless, the first chapter concerns not natural selection, but what he calls artificial selection (much of which is about preposterous fancy pigeons bred for competition), as a means of showing that creatures are mutable over the generations. Galton figured that this principle could be applied to humans.

According to Darwin's model, there has to be variation among individuals in a population for evolution to occur. Some members must be naturally better adapted for success in the struggle for existence in any particular environment. Galton took that idea and applied it very specifically to the huge disparities in intellectual capabilities of man (he expressed little interest in the abilities of women). He believed that these traits were innate to the extent that neither social nor educational environments could overwhelm them. His first scientific book concerned the observation that eminence runs in families. *Hereditary Genius* was published in the United Kingdom in 1869 and lays out an empirical theory of the greatness of British men.

The seeds of Galton's eugenic thinking first came to light in a pair of magazine articles in 1865, but they bloomed in *Hereditary Genius*. It specifically analyzes the inheritance of intellectual ability, framed according to the fairly arbitrary judgments of who Galton decrees are great men. Their eminence is judged via reputation and status, garnered from obituaries and biographies; by Galton's reasoning, the truly eminent men number only 250 in a million (or 1 in 4000). The concept of genes did not yet exist in 1869, or indeed any mechanism for biological heredity, but the success of genius men in Galton's eyes showcases an innate and heritable condition, passed from father to son, and maintained within important families: "a man's natural abilities are derived by inheritance, under exactly the same limitations as are the form and physical features of the whole organic world."

This, of course, is empirically true: we are biological creatures, an ape evolved from extinct apes, and cousin to living apes. We are a symphony of what is innate and how that plays out in our lived lives—nature and nurture. It was Francis Galton who later coined the phrase "nature versus nurture," and in his appraisal of eminence, he very clearly favors the former.*

Hereditary Genius is a strange, awkward book. In the late twentieth century, academic historians and teachers began moving away from "great men theory," where history could be told via the intrepid deeds of unique men of pivotal influence. The Scottish philosopher Thomas Carlyle developed this view, asserting in essays, books and lectures in the 1840s that it was in the innate abilities, heroic acts or divine inspiration of individuals—always men—that history could be explained. In Galton's work, this idea reaches its apex. He insists upon the superiority and rarity of men of true eminence, from the "Judges of England from 1660 to 1868, the Statesmen of the time of George III., and the Premiers [Prime Ministers] during the last 100 years" as well as "Commanders, men of Literature and of Science, Poets, Painters, and Musicians."

Galton would later emerge as a founding father of statistics, and a de facto father of genetics: it is from his biometric analysis of people that much of the modern study of heredity, families and sex derives, and the mathematical fusion of heredity with Darwin's theory of evolution by natural selection would occur under his auspices. *Hereditary Genius* differs from other works of "great men" history because of its methodological approach. Galton strives to use statistics to make his case for the rarity of

---

* Possibly influenced by Prospero's lament for Caliban in *The Tempest*: "a born devil on whose nature Nurture can never stick."

eminence, and its hereditary qualities. As with his beauty map, his aim is to quantify something ill defined and vague.

Over the next fifty years or so, as eugenics developed into what was then regarded as a science, the ability to measure the qualities—or indeed *the* quality—of a person was essential for the design of eugenic programs. This would involve different metrics over the years, but none more than the intelligence quotient in its various forms—IQ. In time, and primarily in America, IQ would be used to categorize people and their intellectual ability such that they might be sterilized, or so that immigrants might be sent back to where they came from. In *Scientific American* in 1915, a public health official named Howard Knox wrote, in language familiar to that of the contemporary eugenicist, "The purpose of our mental measuring scale at Ellis Island is the sorting out of those immigrants who may, because of their mental makeup, become a burden to the State or who may produce offspring that will require care in prisons, asylums, or other institutions."

Galton's metrics were far less rigorous than even the early forms of standardized IQ tests that were the tools of U.S. immigration policy in the first few decades of the twentieth century. He introduces a ranking for classes of people, A–G, and observes the rarity of As as well as Gs according to a statistical technique invented by the Belgian astronomer and demographer Adolphe Quételet—the law of deviation from an average. From that Galton calculates that if there are sixteen grades of men, eight above average, that only the sixth, seventh and eighth top grades will rank as eminent:

> We have seen in p. 25, that there are 400 idiots and imbe-
> ciles, to every million of persons living in this country;
> but that 30 per cent of their number, appear to be light
> cases, to whom the name of idiot is inappropriate. There
> will remain 280 true idiots and imbeciles, to every million
> of our population.

The precision of these numbers is suspect: 250 first-class men per million, 280 true idiots and imbeciles, everyone else somewhere in the middle, in a way that anticipates the statistical distribution known as the normal distribution, sometimes called a bell curve.

Galton was a hoarder of information, a data junkie, whose mantra was "whenever you can, count." He saw the power of science, statistics and empirical arguments in fomenting political change. Nevertheless, there is a notable tension in *Hereditary Genius* and his later works.

The metrics he applies, in an era before IQ tests for cognitive abilities, are deeply questionable and reveal the inherent prejudice of men of power: they include the results achieved in university exams, and rest upon the preeminence of Oxford and Cambridge universities, as well as the major exclusive private schools where men of hereditary power typically went. He asserts that kinship transmits a higher degree of eminence, and that men do so more readily than women, though he admits that the sample size is low—a factor that continues to hamper the quality of published science to this day: "The numbers are too small to warrant any very decided conclusion; but they go far to prove that the female influence is inferior to that of the male in conveying ability."

This is a revealing point and worth noting: Galton was working in the days before the molecular means of inheritance had been revealed. The scientist-friar Gregor Mendel was pottering away breeding his pea plants in Moravia at exactly the same time, but his results—which gave us the concept of the gene as a unit of inheritance—would not be brought to light until the Victorian era was over. We now know the exact mechanisms by which women and men pass on their genetic material, via sperm and egg, and the ways in which their genomes are halved and shuffled as those cells are born, before coming together to form a whole and unique genome at conception. The contribution is not exactly equal though. Due to the minuscule size of the male Y chromosome (passed from fathers to sons), and the addition of the separate mitochondrial genome, which sits not in the nucleus but in the egg's plasm and is therefore passed only from mothers to children, women contribute fractionally *more* genetic material to their children than fathers. Galton was not to know that, but it shows that his model was wrong. He assumed a sexed biological role of inheritance that was not there. Ultimately, nurture would explain the discrepancy far more than nature.

In the end, his thesis is hamstrung by its *lack* of objectively rigorous data. Galton blithely acknowledges this—it is the nebulous concept of reputation on which his whole house of cards sits. The qualities of his geniuses are determined by the "opinion of contemporaries, revised by posterity—the favourable result of a critical analysis of each man's character, by many biographers." His data is literally opinion. *Hereditary Genius* is a superlative showcasing of confirmation bias. Galton asserts the greatness of his geniuses, and at great length explains that they are geniuses because they are great.

Hereditary Genius is also a fundamentally racist thesis, in a way that was typical, though not universal, for Galton's time and place. Galton's racism was deep, consistent and robust, even for his era. It was explicitly White supremacist:

> [T]he negro race is by no means wholly deficient in men capable of becoming good factors, thriving merchants, and otherwise considerably raised above the average of whites—that is to say, it can not unfrequently supply men corresponding to our class C, or even D. It will be recollected that C implies a selection of 1 in 16, or somewhat more than the natural abilities possessed by average foremen of common juries, and that D is as 1 in 64—a degree of ability that is sure to make a man successful in life. In short, classes E and F of the negro may roughly be considered as the equivalent of our C and D—a result which again points to the conclusion, that the average intellectual standard of the negro race is some two grades below our own.

His grading of people goes on:

> The Australian type is at least one grade below the African negro. . . .

[T]he number among the negroes of those whom we should call half-witted men is very large. Every book alluding to negro servants in America is full of instances. . . .

The average standard of the Lowland Scotch and the English North-country men is decidedly a fraction of a grade superior to that of the ordinary English.

The ablest race of whom history bears record is unquestionably the ancient Greek, partly because their masterpieces in the principal departments of intellectual activity are still unsurpassed, and in many respects unequalled, and partly because the population that gave birth to the creators of those master-pieces was very small.

Note the casualness of his rankings. Also note the adoration of classical civilizations—a recurring theme for obsessive rankers of people. Though Galton and his intellectual descendants wanted this to be a science, informed by statistical analysis and rigor, it is not and cannot be value free. The idea that lives should be as free from pain and suffering as possible is uncontroversial, but these analyses of a vague sense of quality of people instead invoke the idea that some people are demonstrably better than others over and over again. There are a handful of geniuses and imbeciles at the two poles of society but in the morass in between, there is also a hierarchy, and the aim of society should be to fill future generations with better people than are there currently. Galton lambasts the Church for imposing celibacy on the great men of the Middle Ages which caused the "moral deterioration" of the United Kingdom, and he castigates

academia for continuing to discourage marriage among Fellows
at British universities, a practice he dramatically likens to religious persecution.

Galton's systematic grading of humans is the backbone of
his agricultural view of humans in society. We are animals, and
animals can be bred. This is central to the eugenics project. Back
in the 1870s, with this clear vision of breeding, the emergence of
improvement in populations becomes Galton's political dogma
for change:

> I argue that, as a new race can be obtained in animals and
> plants, and can be raised to so great a degree of purity that
> it will maintain itself, with moderate care in preventing
> the more faulty members of the flock from breeding, so a
> race of gifted men might be obtained, under exactly sim-
> ilar conditions.

This expression of a Darwinian view of humans sets out the
principle of positive eugenics—that is, the improvement of
future generations' quality (however arbitrary that definition
is) by selective breeding. People can be better, and nations can
comprise a better stock if efforts are made to encourage good
breeding of right-minded people, as Plato had suggested two
thousand years earlier. Without that nudging of society in the
right direction, we drift toward degeneration with the many
babies of the unfit swamping those who achieve eminence.

Galton's ideas—and the eugenics movement that they
spawned—are simultaneously radical and traditional. With

the new science of evolution applied to humans, society must escape the doldrums in which it has languished, but that change is engineered to maintain Great Britain's dominance over the world and harks back to the classical notion of a lost golden age. We must change in order to stay the same. No matter how scientific Galton tried to be, his ideas emerge from a fear of the decline of civilization.

Weird, fascinating, flawed, racist, sexist and blindly biased though it is, *Hereditary Genius* fires the starting pistol for a methodological and (pseudo-)scientific endeavor never seen before in the long history of population control. Before the word existed, eugenics is born, and Galton its midwife, possessed by a religious conviction that only statistics will save his brethren: "When the desired fulness of information shall have been acquired then, and not till then, will be the fit moment to proclaim a 'Jehad,' or Holy War against customs and prejudices that impair the physical and moral qualities of our race."

GALTON'S INFLUENCE

Reviews of *Hereditary Genius* were mixed. Some were clearly unpersuaded by Galton's reduction of the social influences of achievement over the genetic. Religious reviewers were angered by his attacks on the Church, and its diminution of Christian charity as a means of improvement. The magazine *Catholic World* began its anonymous review: "Mr. Galton is what in these days is called a scientist, whose pretension is to assert nothing but positive facts ... ascertained by careful observation and experiment and induction therefrom . . ."—which may well be intended as a haughty sleight, but to most scientists' ears today will sound both accurate and desirable. Some reviews, notably scientific ones, commended its data-driven approach. The naturalist Alfred Russel Wallace—who came up with a correct but slight version of natural selection at the same time as Darwin a decade earlier—reviewed the book for the scientific journal *Nature* and concluded that it was "an important and valuable addition to the science of human nature." Other reviewers were alert to the arbitrariness of Galton's definitions of eminence, and his failure to acknowledge professional trades—notably the law—that promoted average men via nepotistic family connections. A review in the London literary magazine *The Athenaeum* declared that it was a sturdy example of the maxim that "anything can be proved with statistics."

None of these arguments has ever really gone away—which is dominant, our inherent nature or how we are nurtured? Untangling the relative influence of DNA and the environment remains one of the great challenges of human genetics in the twenty-first century. We continue to rely on twin studies— invented by Galton—as a means of canceling out the genetic influence on individuals and revealing the environmental, as identical twins have near identical DNA, and so any difference between them should be attributable to nurture, not nature. Statistical analyses of genomic data—the accumulated mass of which is now galactically vast—are often misinterpreted, misunderstood and misrepresented. So-called hereditarians are a vocal but mostly fringe collective of online commentators, activists and the occasional scientist who seek to amplify the role of genetics over that of the social environment for many personality traits, particularly in relation to race and intelligence, sometimes with misunderstood, weak or even fraudulent data.* Galton was the first hereditarian.

Support from scientists was enormously significant in normalizing Galton's ideas of population control. The journal *Nature* was then brand-new but was rapidly becoming an influential pillar of the scientific establishment, featuring scientists of the very highest caliber (for example, Darwin would publish his last ever writing there in 1882, on barnacles, informed by the work of the grandfather of Francis Crick). The journal's editorial stance was to immediately advocate for governmental intervention in changing breeding patterns for British people, with the lead article in December 1869 suggesting that:

we may vary the circumstance of life by judicious legislation, and still more easily by judicious non-legislation,

---

* Some frequently used data sets selected by the controversial intelligence researcher Richard Lynn that purport to represent national IQs—including those that claim to show many sub-Saharan African countries having significantly lower averages than European countries—are meaningless, invalid, and absurd. For example, the national IQ for Botswana in the NIQ_QNWSAS data set is 69.5, but upon inspection, this number was arrived at with a single sample of 104 natively Tswana-speaking high-school students aged seventeen to twenty, tested in English, and thus not comparable with that of other countries. For Somalia (measured at IQ 67.7): a single sample of refugees aged eight to eighteen tested in a Kenyan refugee camp; and so on. Nevertheless, these data are unquestioningly cited as valid by dozens if not hundreds of peer-reviewed academic papers, despite their fatal flaws.

so as to multiply the conditions favourable to the development of a higher type; and by the same means we may also encourage, or at least abstain from discouraging, the perpetuation of the species by the most exalted individuals for the time being to be found. Parliament, being an assembly about as devoid of any scientific insight as a body of educated men could possibly be, has not as yet consciously legislated with a view to the improvement of the English type of character.

Two years later, in his second masterpiece, Darwin addressed the role of evolution in the history and future of humankind. *The Descent of Man* is a tremendous book that also contains many more problematic ideas than his previous works, not least because it showcases several of Darwin's views that were sexist and racist in a manner in keeping with the times, even though he was a liberal abolitionist, and politically a long stretch from his cousin. Eugenic ideas are present, though not a major focus:

With savages, the weak in body or mind are soon eliminated; and those that survive commonly exhibit a vigorous state of health. We civilised men, on the other hand, do our utmost to check the process of elimination; we build asylums for the imbecile, the maimed, and the sick; we institute poor-laws; and our medical men exert their utmost skill to save the life of every one to the last moment. There is reason to believe that vaccination has preserved thousands, who from a weak constitution would formerly have succumbed to small-pox. Thus the weak members of civilised societies propagate their kind. No one who has attended to the breeding of domestic animals will doubt that this must be highly injurious to the race of man.

At this time, not only were the mechanisms of inheritance— how genes are passed from parent to child—unknown, but so were many of the basic patterns of heredity. Though in *The Origin of Species*, Darwin had outlined the correct view of how evolution works on a macro scale, natural selection had temporarily fallen out of favor. Darwin himself was drawn to a view that is much more in line with Lamarckian inheritance—the now unfairly derided theory that traits modified during life would be passed on to their offspring.* After Darwin died in 1882, various other theories about how traits were passed on circulated, though the actual mechanisms of sexual inheritance were far from settled. No one could yet see precisely how certain traits ran through families, and how patterns of inheritance flowed down generations.

---

* Larmarckian inheritance is sometimes mocked for its stupidity. We now know that the giraffe did not get a long neck because it strives to get the juiciest leaves, and the offspring of blacksmiths don't have bigger muscles because their father smashed iron all his life. Evolutionary change does not act on characteristics that are acquired during life. Lamarck was a good scientist and his ideas do not deserve to be mocked. He was wrong, and scientists should always be very happy to be wrong.

But at least they were talking about it. As the broader idea spread, others were beginning to discuss the mechanisms of birth control. The British suffragist and philosopher Jane Hume Clapperton expressed a view that became common among suffragists and feminist campaigners a few years later as a means of improving the lives of women: the suggestion that birth control and sterilization would be better for improving the British stock. This period sees the rise of sex education and emancipation of women, and many first-wave feminists embraced eugenics as a means of improving the opportunities and quality of life for not just women, but also people more generally.

The senior politician and science enthusiast George Campbell, eighth Duke of Argyll, clearly not a romantic, confidently declared in 1886 that:

> we have enough physiological knowledge to effect a vast improvement in the pairings of individuals of the same or allied races if we could only apply that knowledge to make fitting marriages, instead of giving way to foolish ideas about love and the tastes of young people.

The science fiction writer Grant Allen—a big supporter of evolution in the late nineteenth century—even went so far as to suggest that women should be emancipated, educated and free to choose the best fathers for their children, which is essentially how sexual selection works in nonhuman animals. And to guarantee success, they should reject monogamy and embrace a series of men to maximize their children's potential. Alfred Russel Wallace, and I expect many men of the time, thought this detestable.*

A POLITICAL SCIENCE

The conversation about population control in the United Kingdom at this point was largely focused on betterment of the stock of a people, rather than pruning the unwanted. But the idea of positive eugenics cannot exist without its opposite. Selecting *for* desirable traits must mean that other traits are less desirable, and are therefore being selected *against*. The existence of desirable people implies that undesirables also exist. You cannot rank things high without there being a lower order, and you cannot only select for enriching the top without at an absolute minimum acknowledging that the people at the bottom will not continue into future generations—it is implicit in positive eugenics that there must also be deliberate selection against the unfit, the undesirable or the defective.

In Victorian Britain, the presence of the undesirable was

---

* During the writing of this book, I was continually bewildered by the recapitulations of history in the present, even the repetition of the same language. In July 2021, as I was finishing a first draft, the South African government began considering a revision to their marriage laws—which permit polygamy—to specifically include polyandry as well. Commenting on the backlash that this proposal engendered, an academic who specializes in polygamy named Collis Machoko told the BBC: "African societies are not ready for true equality. We don't know what to do with women we cannot control."

being redefined. Expanding working classes, urbanization and visible poverty were all growing concerns for anxious middle-class society, and the gradual shift away from the charity of Tudor poor laws was passing the responsibility of dealing with those assumed to be less well equipped to handle the hardships of life from the Church to the state. The lunacy industry was developing rapidly, and government-sanctioned institutes were commonplace. The Madhouses Act of 1774 had been replaced by the Lunatic Asylums Act in 1853, which saw the building of more than forty institutions to detain and house a bewilderingly diverse range of people. The reasons for admission were vast and loose. They tended to be stratified by class—working-class people were far more likely to be incarcerated than those in wealthier classes for the same conditions—and targeted women more often than men. Asylums housed children and adults, violent criminals, epileptics, people with tuberculosis and other diseases, alcoholics and women with menstrual problems. All these conditions would eventually be the targets of eugenic purification in Germany, America and elsewhere.

"Lunatic" was something of a broad catch-all term for people to be hidden from society. Part of the model of Victorian rehabilitation was a softer approach than the eighteenth century, with prettified gardens and caring convalescence at least for the more middle-class maladies. But the second half of the nineteenth century saw a general move toward the brutal and clinical. More distinctions were being introduced, each with its own pseudopsychiatric labels—low- and mid-grade idiots, the feebleminded, imbeciles, and other terms that seem almost comically vague now, but harbored the diagnostic criteria that could result in lifelong internment in Victorian Britain.

This is the landscape in which the ideas of positive and negative eugenics took hold: a changing demographic structure of the country, with a ruling powerful elite seeking improvement of their power, or at least the maintenance of it in the face of a growing and visible working class, expanding with immigration.

Following Galton's *Hereditary Genius*, and other works discussing the sculpting of populations, many books, articles and pamphlets by intellectual heavyweights of the era over the next few years placed eugenics at the center of popular discourse in British life. Galton's reputation and standing in society would only grow, with the publication of several books and papers over the next four decades that refined and developed theoretical and practical eugenics.

The powerful really only seek one thing, and that is to maintain their power. Galton was working in an era of expanding colonies, but *after* slavery was formally outlawed across the British empire. The bounteous financial fruits of colonialization were continuing to fuel industrial development back home, which meant growing cities and a flourishing working class, but also global migration, visible extreme poverty and, with it, disease and criminality. Furthermore, insurgency and revolt were commonplace in the territories occupied by British colonists, often but not exclusively on the part of the people who were actually born there. The Boer Wars were raging in southern Africa, and the military might of the British was being successfully challenged, which prompted the formation of the Committee on Physical Deterioration in 1903. The United Kingdom's principal African colonizer Cecil Rhodes, Francis Galton, a young Winston Churchill and many other leaders were open in expressing the sense that White supremacy was the moral duty of the British: to fill the world with the dominant race of the best type of men.

This era—the last decade of the nineteenth century and first of the twentieth—provided a rich soil for the formalization of eugenics. The idea of racial decay was the fertilizer. Superiority could be achieved with science as the engine of social change. By now old and august, Galton proposed eugenics be treated with the energy and fervor normally accorded to religion: a scientific creed could replace traditional faith. Toward the very end of his life in 1910, he even did something that is generally ill advised for scientists trying to promote their ideas: he wrote a novel.

In *The Eugenic College of Kantsaywhere*, the people of the titular country "think more of the race than of the individual." As in Plato's utopian *Republic*, the government of Kantsaywhere directs the quality of its people by controlling their sex lives. Prospective parents are screened with physical and psychometric tests, and either given the go-ahead to get on with it, or face banishment to labor colonies with enforced celibacy. The novel was never published, and was suppressed and partially destroyed after Galton's death by his sister, probably because she was concerned about the appropriateness of the sex scenes for polite society.* In Galton's otherwise lumpen prose, he becomes excitable with detailed descriptions of the assessments of couples, which include his lifelong obsessions: ancestry, literary and aesthetic insight, and biometrics. These reflected his meticulous collation of immense tranches of personal and biometric data acquired in multiple ways over many years, notably when 9,337 people paid him to submit their vital statistics—height, weight, arm span, visual acuity, punch strength—at the 1884 International Health Exhibition in South Kensington. Getting average joes to shell out three pence (around one dollar in today's money) to hand over their data was a brilliant move. Galton wanted the numbers—we call it Big Data today—and got people to *pay* to give it to him, in exchange for a scientific personal trinket: a carbon-copied card filled out with those vital statistics. It's a ruse mirrored today by commercial genetic companies such as 23andMe, who provide you with ancestry and genetic information in exchange for your personal genomic data, which you pay to give to them, which they can then use (unless you opt out) to develop drugs to be sold back to the public.

Unlike his many works of nonfiction, *The Eugenic College of Kantsaywhere* never saw the light of day, but the ideas that Galton had ignited were becoming mainstream. The normalization of eugenics may well seem baffling to us today. The idea that class and ability taxa were both inherent to society and could be controlled by limiting the freedoms of human beings

* What remains of the manuscript is available online on the University College London Library Special Collections webpages.

is a view that is generally alien to Western values in the twenty-first century.* Others took the idea to grotesque lengths. In 1908, as eugenics fever was hitting religious ferocity, a young D. H. Lawrence wrote to the children's author Blanche Jennings Thompson:

> If I had my way, I would build a lethal chamber as big as the Crystal Palace, with a military band playing softly, and a Cinematograph working brightly; then I'd go out into the back streets and the main streets and bring them all in, the sick, the halt and the maimed: I would lead them gently, and they would smile me a weary thanks; and the brass band would softly bubble out the Hallelujah Chorus.

A lethal chamber. Of course, Lawrence's psychotic fantasy would be realized eventually, not with gentle music and a soothing movie, but in cattle trucks, with Zyklon B and tattooed arms.

---

* Generally, though not totally. During his political career, Donald Trump was not shy of retching up opinions that capriciously lurched between poles of ignorance and lies, but one view that he has expressed repeatedly is that some people, and predictably himself, have "good genes," and these are selected and preserved in families as with Thoroughbred racehorses. This model was inherent to both American and Nazi eugenics policies in the 1930s. "You have good genes, you know that, right?" he burbled moronically in September 2020 to a rally in Minnesota—a state that is more than 80 percent White. "You have good genes. A lot of it is about the genes, isn't it, don't you believe? The racehorse theory. You think we're so different? You have good genes in Minnesota." That, dear reader, is straight-up old-school eugenics.

COMING TO AMERICA

What started with the sciencification of population control in the 1860s was becoming not only mainstream but international too. The same but different circumstances—mass immigration, ongoing military conflict, a visible and growing underclass—were present in many Western countries, where similar conversations about how to fix society with this new science were taking place. The United States and Canada were both warmly receptive to the ideas of eugenics, and would go on to embrace them like no other nations. In doing so, they directly reciprocated eugenics ideas and policy and influenced growing powers in Europe, both British and German. As with most history, there is no simple linear narrative here, but the route to the Holocaust, and with it the reversal of the popularity of eugenics, can be traced from Britain via North America to the Third Reich. In the reckoning of Nazi scientists and doctors in the Nuremberg trials, the war criminals who enacted eugenics with industrial efficiency—which we will come to in a few pages—cited the influence and policies of America as part of their inspiration.

As in Britain, widespread support for the principles of eugenics was enjoyed across political and class divides. Titans of American industry with virtually limitless wealth plowed their millions into eugenics projects. The formal and scientific base came with the founding of the Eugenics Record Office (ERO) in Cold Spring Harbor in Long Island in 1910. Over the years, the majority of its funding would come from the philanthropy of the Carnegie Institution for Science, the Rockefeller Foundation and from the estate of Mary Harriman, the multimillionaire philanthropist and widow of railroad tycoon E. H. Harriman.

Just as the Eugenics Laboratory at UCL would evolve into the genetics department (of which I am a member), the ERO was a department that was founded within the Cold Spring Harbor Laboratory, today a global powerhouse of genetics research, and hub for the Human Genome Project in the late twentieth century. In 2018, it was rated by *Nature* as the single best research institute in the world.

The key person in the ERO's inception, and indeed the central figure for eugenics in the United States, was Charles Davenport. Like Galton and his disciples in the neonate genetics labs of London and Cambridge, Davenport was the driving force for the development of studying human heredity, and applying it as an ideology. There are few men more deeply embedded in the fabric of American eugenics.

Davenport was raised a New Yorker, the son of strict Protestant parents, and claimed American ancestry back to the seventeenth century. His father was authoritarian by nature, and had him home-schooled in Brooklyn Heights. Nevertheless, at the age of twenty-one he entered Harvard to study biology. He completed a PhD in zoology in 1892 and then immediately began a seven-year stint there as an instructor. From there Davenport's scientific star rose and rose, first to an assistant professorship at the University of Chicago, and then to become the director of the Station for Experimental Evolution at Cold Spring Harbor, funded by the Carnegie Institution. There, he was known as a sound breeder of animals in service of testing Gregor Mendel's

patterns of inheritance, the ones that had only been rediscov-
ered in the previous few years, but he also had a reputation for
being shoddy, superficial and rushed in his subsequent analysis.

Davenport had met Galton in London in 1899. Like the
Englishman, the American devoted his life to studying first
human heredity and later its politicized bastard child, eugenics. Like Galton, Davenport saw explanations for human heredity in statistical analysis, and enthusiastically and uncritically
embraced Mendel's laws as the rigid model of heredity. But also
like Galton, Davenport saw the social change of the American
people as the manifest destiny of this new science. His interests moved from animal heredity to the traits of humans, and
though he maintained an interest in physical traits and diseases,
before long he began pontificating on the standard canards of
the Victorian eugenicist and social campaigner: alcoholism,
criminality, feeblemindedness, intelligence, manic depression
(and, weirdly, seafaringness). Heredity, he wrote in 1910, "stands
as the one great hope of the human race; its savior from imbecility, poverty, disease, immorality." Davenport's adherence to
Galtonian thinking also bound him to a belief that all of these
undesirable characteristics were largely genetic—nature not
nurture. In 1913, he took on prostitution as a social ill, and found
that out of 350 sex workers all bar one had taken to this profession because of "innate eroticism" rather than socioeconomic
necessity.

In 1910 Davenport established the ERO with a $35,000 per
annum donation from Mary Harriman. It had two objectives:
to campaign for the deployment of eugenics policies around the

United States—specifically, for reduced immigration and the forced sterilization of "defectives"—and, principally, to compile records of families and their characters, as if constructing a national pedigree to be pored over, and from which the American stock could be improved. Volunteers collected family data via questionnaires, and compiled data sets from which Davenport and his scientists would extract information about the patterns of inheritance of a phalanx of human traits. This was crucial to establishing who should be preserved and who should be expunged. Sometimes, these polls were sent out under the ruse that they would help people to understand heredity and ancestry in their own families, while the harvested information would be of scientific use for the eugenicists at the ERO, again reminiscent of the current fashion for paying customers seeking out trivial ancestry infodumps via commercial DNA testing companies such as 23andMe, when the customer's genomic data becomes their property.

For a country founded on principles of freedom, turn of-the-century America was revving up the restrictions on personal reproductive liberties, but eugenics policies were already in place before then. In 1896, Connecticut legislated against the marriage of epileptics, imbeciles, habitual drunkards and the feebleminded, with Kansas, New Jersey, Ohio and Michigan soon following suit. Under these laws, both husband and wife typically had to present certificates of medical soundness before the marriage could take place. In Indiana in 1907, the first official sterilization program was legally enshrined. By 1931, thirty states had performed the sterilization of over twelve thousand men (via vasectomy) and women (via fallopian tube severing). Half of these were in California.

AMERICAN HISTORY X

Eugenics had become fundamentally tethered to a threat. The infanticide of ancient cultures had always been about pruning the weak, deformed or unwanted, and eugenics principles had initially emerged first for the enhancement of positive traits in a population. But negative and positive eugenic practices are implicitly linked and cannot really be isolated from each other. Furthermore, the imagined peril was sharpened by mass immigration. People coming from abroad would dilute the purity of the White race, weakening it. The threat was most explicitly expressed by Charles Davenport: "New blood will make the American population darker in pigmentation, smaller in stature, more mercurial . . . more given to crimes of larceny, kidnapping, assault, murder, rape, and sex-immorality."

I wonder if this threat is felt more acutely in young countries such as the United States, with an underlying reluctance to publicly acknowledge that this is precisely what colonizers had done to millions of Indigenous people. Estimates vary but following the European invasion of the Americas and the genocide of Indigenous people that started with Columbus, around 10 percent of the world's population was killed. My father was raised in the 1950s in New Zealand, arguably one of the better examples of colonizers and their relationship with Indigenous people. Nevertheless, he was taught that the Māori were not the first people on those islands. Instead, the Māori had violently replaced the Moriori—an earlier and more primitive native tribe. This was a fabrication perpetrated by nineteenth-century British–New Zealander writers to justify their own ongoing conquest. Perhaps ideas of population replacement strike harder in uneasy countries whose recent ancestors were the successful perpetrators.

It's an idea presented again and again throughout history, but perhaps never more conspicuously than in the time of eugenics. It's right there, explicitly in the first few pages of F. Scott Fitzgerald's magnificent and most famous novel, *The Great Gatsby* (1925). Daisy's violent and cruel husband, Tom Buchanan, frets a refrain that can be heard in the echo chambers of the racist internet today: "if we don't look out the white race will be—will be utterly submerged. It's all scientific stuff; it's been proved."

Well, it hadn't been proved, and never has had much merit, scientifically or otherwise. Scott Fitzgerald had met Mary Harriman Rumsey, the eldest daughter of Mary and E. H., at Old Westbury, which was a significant model for the Long Island residence of Jay Gatsby and his cohort of careless people. Untold wealth, lavish parties and more than a passing interest in the new science of eugenics was the lifestyle of these American aristocrats: Mary was nicknamed Eugenia by her Barnard College classmates, due to her fascination with the idea of scientific betterment for the American people. In Fitzgerald's notebooks, he makes references to attending her parties in the 1920s, as does his wife, Zelda, in a 1930 letter. These upper-class Americans were Fitzgerald's templates for the cast of *Gatsby*, and eugenics was a key part of the discourse in that rampant social scene. It was via Mary Harriman Rumsey that Charles Davenport sought funding from her mother for the ERO in 1910.

In *Gatsby*, Tom Buchanan cites a book called *The Rise of the Colored Empires* by a fictional scientist called Goddard. It's a very thinly disguised reference to three of the major American eugenicists of the era. One is Henry H. Goddard, whom we will come to in just a minute. The title riffs off *The Rising Tide of Color: The Threat Against White World-Supremacy*, a bestseller written in 1920 by T. Lothrop Stoddard, a member of the Committee on the Heredity of the Feeble Minded under Charles Davenport at the ERO. And perhaps Fitzgerald got from Stoddard to Goddard by splicing in a G from Madison Grant, who wrote the introduction to *The Rising Tide of Color*. Grant had also written a 1916 book called *The Passing of the Great Race*, proclaiming the so-called Nordic theory, a treatise of pseudoscientific White superiority that asserted a hierarchy of the world's peoples drawn from anthropological theories of the time. This model asserted, erroneously, that there were three types of humans: Caucasoid, Mongoloid and Negroid. Of the Caucasoids, northern Europeans from Scandinavia, northern England and Scotland, and Germany were unequivocally the best. Its arguments are seriously unhinged. Grant asserts that according to measurements of busts of their heads, Leonardo da Vinci, Michelangelo and Dante were clearly of Nordic extraction. He fulfills the adoration of the classical world trope with a crazy made-up theory that tall blonde Nordic men seeded the great cultures of Rome, Greece and Egypt before retiring back to their northern homelands.

But despite their inherent dominance, Grant contemplates the "racial suicide" of Nordic people, their purity sullied—"mongrelized"—by mating with lesser humans, and the persistent eugenics theme of being outbred and outnumbered by them as well. *The Passing of the Great Race* was a profoundly influential screed. Adolf Hitler described it as his "Bible," and with the easy substitution of "Aryan" for "Nordic," it became pivotal in developing the eugenics policies of racial hygiene in Nazi Germany. It persists like a turd that won't flush as an influence in antisemitic conspiracies today: interracial marriage with Black people, Hispanics and Muslims, coupled with low fertility rates among White people, and rights for immigrants and minority groups surely will spell the end of the White race in America, and it's all being orchestrated by an invisible Jewish global elite. As we've seen, "the Jews will not replace us" was the chant of the tiki-torch-wielding racists in Charlottesville in 2017.

The theme in books both real and fictional is a muscular victimhood of the powerful. It runs through the eugenic thinking of that age: Buchanan, Goddard and Stoddard reflect the paradoxical view that "we"—meaning the White, Western and wealthy—are naturally in charge, but "our" power is waning as time marches on, and it must be protected by whatever means necessary.

As mentioned above, the third of Fitzgerald's probable influencers was Henry Goddard, a psychologist who specialized in working with people with mental deficiencies in New Jersey, at the Training School at Vineland. It's an institution that still exists, nowadays to promote independent living for people

with learning disabilities. Goddard was an early adopter of the
standard IQ test, which had been developed in France by the
psychologists Alfred Binet and Théodore Simon in 1905. Goddard was the first person to translate it into English for use in
America.

Goddard used IQ to generate specific delineations for psychological evaluation: he introduced the term "moron" to
describe people scoring between 51 and 70, and used "imbecile" for those in the 26–50 range, and "idiot" for those below
26. These standards were applied for years, and those categories were used as eligibility criteria for sterilization. IQ, for the
record, is a perfectly valid test for measuring cognitive ability.
It has been studied extensively for more than a century, and we
have colossal data that is consistent and has predictive power for
life outcomes, including educational level, longevity and professional success. Like all metrics that provide a single number as
a measure for a complex set of traits, it is not absolute and is not
without flaws. The oft heard complaints about IQ include that it
only measures how good you are at IQ tests and that it is culturally specific and therefore inherently biased. These are true and
known by all psychologists. But these valid criticisms do not
invalidate IQ as a useful scientific tool.

Of course, the deployment of IQ testing as a useful and absolute metric that would determine who gets to remain free and
fertile was abhorrent. There is an ongoing controversy today
about the validity of IQ in psychological research and genetics,
and with sound reason; these spurious actions of early eugenicists may well be the source of the animosity.

As in Britain, American names associated with reproductive rights for women were also key players in the promulgation of eugenics. Like many of her wealthy or influential peers, Margaret Sanger, who thought the well-to-do were not having enough children and the poor were having too many, became one of the first pioneers of reproductive autonomy for women. She popularized the phrase "birth control" and opened the first clinic dealing specifically with reproduction in women in New York in 1916. Five years later, she launched the American Birth Control League, which evolved into the Planned Parenthood Federation of America in 1942. Sanger was also an active member of the American Eugenics Society and was the editor of the *Birth Control Review*: "More Children for the Fit. Less for the Unfit" ran the cover in May 1919. Those who qualified as unfit were not only people with disabilities, but those of lower intelligence and criminals, and her plan was the gradual suppression, elimination and eventual extinction of defective stocks—"those human weeds which threaten the blooming of the finest flowers of American civilization." Sanger proposed that the unfit should require official permission to have babies—"as immigrants have to apply for visas."

Again, it's hard for us to imagine quite how mainstream these ideas were. The fretting of the wealthy and powerful about a threatening underclass was also successfully passed along to working-class families, particularly in rural America, ironically as a response to the degeneracy and vice of the Roaring Twenties. "Fitter Family" competitions were popular at state fairs and agricultural festivals, where almost exclusively White Protestant families would be selected and assessed for their quality. The first of these was at the Kansas Free Fair in 1920, organized by Mary T. Watts, of the Iowa Parent-Teacher Association, and Dr. Florence Brown Sherbon, a child welfare officer from Kansas. They sought and received the support of Charles Davenport of the ERO and the American Eugenics Society. Their aim was to promote traditional family values and large nuclear families as bulwarks against the degeneracy of lesser races. The agricultural setting was no coincidence. Watts was explicit in drawing a direct comparison for farming folk: "while the stock judges are testing the Holsteins, Jerseys, and Whitefaces in the stock pavilion, we are judging the Joneses, Smiths, and the Johnsons."

Families were medically examined, quizzed about their health record, personality and temperament, had their IQs tested, and blood and urine taken. The winners received a medallion, inscribed with the legend from Psalms 16:6: "Yea, I have a goodly heritage." Every participant received a personal report, and the ERO kept carbon copies in Cold Spring Harbor.

Another Iowan would have a major role in the legalization of enforced sterilization in both the United States and Germany. Harry Laughlin, Davenport's deputy, was the superintendent of the ERO for eleven years from its inception in 1910. A Princeton graduate and former teacher, he was a eugenics zealot, touring the country to spread the word and to train people in the practice and legal framework being rolled out across America. He published an influential book in 1922, *Eugenical Sterilization in the United States*, which outlined his standard eugenic views, that inferior races were threatening the genotype of White

Americans, and contained a "Model Eugenical Sterilization Law," which purported to standardize the legal framework for involuntary sterilization that had been dribbling out since 1907. Over the following years, more than thirty states used Laughlin's model, and in 1933 it was adopted by the Third Reich just four months after Hitler seized power.

The legislation that grew out of Laughlin's work allowed for the sterilization of some seventy thousand people in America in the twentieth century. The categorization of people who qualified was imprecise, ranging from the "feebleminded" with all its vague qualifications, many derived from spurious IQ tests, to those with physical disabilities, or diseases and conditions that we now know to be largely nongenetic in their origins. It's not easy to disentangle race from eugenics in America, as poverty associates closely with disease burden, and African Americans and Native Americans were (and are) typically among the most disadvantaged socioeconomic group.

Even so, the language of eugenics permeated African American culture too, as leaders in the emancipation of descendants of the enslaved tried to improve their lives. Many Black writers, scholars and families were influenced by and supported some aspects of this intersection of eugenics and public health. Though many of the eugenics policies were aimed at the promotion of White people and the minimization of the Black underclass—most of whom were only one or two generations distant from the enslaved—even the National Association for the Advancement of Colored People (NAACP) advocated for "better babies" as a model for its own continued success. The

most significant Black scholar and activist of this time was
W. E. B. DuBois, a founder of the NAACP, who wrote several
times for Margaret Sanger's *Birth Control Review*, noting that
"among human races and groups, as among vegetables, quality
and not mere quantity really counts." There is a large body of
work analyzing DuBois's scholarship, some of which focuses
on this seeming conundrum—that a policy targeting marginalized people could be adopted to liberate them from cycles of
poverty and an oppression inherited from decades of slavery.
DuBois's views cannot be summarized easily, but in his earlier
years he subscribed to a view that rejected both the inferiority
of Black people and the notion that traits and behaviors were
immutably locked into our biology heritage, and unalterable
by nurture.

Later though, his views evolved to adopt a tone that is at least
superficially reminiscent of the standard tropes of mainstream
eugenics. He wrote in 1932 that educated African Americans
were having fewer children, and this was a problem. And on the
other hand:

> the mass of ignorant Negroes still breed carelessly and
> disastrously, so that the increase among Negroes, even
> more than the increase among whites, is from that part
> of the population least intelligent and fit, and least able to
> rear their children properly.

DuBois felt that African American success was served by better
breeding. He even endorsed the model of the Fitter Family cam-

paigns from rural Kansas, with prize baby contests featuring in the NAACP's monthly magazine *The Crisis*. Part of the motivation for Better Baby competitions was to fund-raise and stand in opposition to the epidemic of lynchings that was happening across the South.

Birth control and the promotion of good breeding were standard tools of eugenics, and were locked in step with the hereditarian view of immutable genetics. But DuBois and other Black leaders who were invested in racial uplift also maintained a staunch grip on the social factors of promoting Black people's livelihood in America, in public classrooms, in clinics and even in the promotion of adoption as a means of raising the qualities of children born into trying conditions. DuBois argued against the impact that certain eugenics practices would have on future generations, including coercive sterilization and bans on interracial marriage.

The anthropologist William Montague Cobb argued later that slavery had already had a positive evolutionary effect on the African American population via "a mass elimination not alone of the weak and unfit, but also of those who were lacking in that individual shrewdness which is a vital essential in self preservation." We now know this to be untrue: in modern genomics we can see where selection has taken place in the genome, by calculating which bits of genes are overrepresented in a population compared with others. A 2014 study analyzed the DNA of 29,141 living African Americans and showed no signs of selection for *any* trait in the time since their ancestors were taken from their African homelands. The average African American genome is different from the average White American genome only because their histories are different.

DuBois was a leading scholar and a champion for civil rights for much of the twentieth century, but as in all cultures, there were a plurality of views. Together with Cobb, we see the adoption of science as a potential solution to social problems. What remains is the cold fact that Black people were disproportionately sterilized under state eugenics policies, because eugenics in a racist society ultimately has hierarchy, racism and classism threaded into its DNA.

THE FEEBLEMINDED BILL

Though the United Kingdom was the birthplace of eugenics, formalization into policy lagged behind the United States, and would never catch up. Francis Galton, Karl Pearson and others had nurtured the public conversation about eugenics, and politicians had embraced their work. Of the many who sought British betterment through eugenics, a young Winston Churchill was one of the most enthusiastic. Churchill was elected to Parliament in 1900, first for the Conservatives, but switched to the Liberals in his first term and remained in that party until 1924. He joined the Cabinet under the Asquith government in 1908, his rapid ascent continuing as he became Home Secretary in 1910.

Churchill was a long-standing and committed racist and eugenicist. Though he led the Allies to great victory against one of the most extreme evils the world has ever seen, his well-documented views on the hierarchies of race make for uncomfortable reading today. The evidence for these charges is overwhelming, but it remains curiously controversial and provocative to point out that one of Britain's greatest heroes had such demonically bigoted views. Perhaps then it is best to air them with his own words, first from 1927:

> I do not admit for instance that a great wrong has been done to the Red Indians of America or the black people of Australia. I do not admit that a wrong has been done to those people by the fact that a stronger race, a higher-grade race or at any rate a more worldly-wise race, to put it that way, has come in and taken their place. I do not admit it.

His racism was persistent through his life too. In reference to Chinese people, in 1954 he commented: "I hate people with slit eyes and pigtails. I don't like the look of them or the smell of them." As Home Secretary though in 1910, his concerns were more parochial and ardently focused on correcting what he perceived as the racial decline of the British. He wrote to Prime Minister Asquith outlining both problem and solution:

> The unnatural and increasingly rapid growth of the Feeble-Minded and Insane classes, coupled as it is with a steady restriction among all the thrifty, energetic and superior stocks, constitutes a national and race danger which it is impossible to exaggerate.... I feel that the

source from which the stream of madness is fed should
be cut off and sealed up before another year has passed.

The Idiots Act of 1886 had attempted to formalize the psychiatric diagnoses of lunatics, idiots and imbeciles, but other
terminology had evolved in the intervening years, as more and
more people were detained into the expanding asylum industry. The report of a royal commission on feeblemindedness was
published in the *British Medical Journal* in 1908, attempting to
quantify the number of citizens who were problems for the state.

Persons who cannot take a part in the struggle of life
owing to mental defect, whether they are described as
lunatics or persons of unsound mind, idiots, imbeciles,
feeble-minded, or other wise [sic], should be afforded by
the State such special protection as may be suited to their
needs.

The report put this number at a suspiciously precise 271,607.
The blurring of the line between psychiatric diagnoses and anti-
social behaviors is clear in this report. It states that 60 to 70
percent of "habitual inebriates" are mentally defective. It is virtually impossible to understand precisely what that diagnosis
means in twenty-first-century terms.

Churchill's correspondence on matters of eugenics is voluminous, and all indicate a very strong commitment to the Galtonian idea of innate undesirable traits, and the prevention of
the unfit from further polluting the British stock. He informed

the prime minister in 1910 that the feebleminded at large in our midst deserved "all that could be done for them by a Christian and scientific civilization now that they are in the world" but should be "segregated under proper conditions so that their curse died with them and was not transmitted to future generations." There is Galton's link between the innate and behavior, and between the science of heredity and politics.

Churchill had read a pamphlet by H. C. Sharp, an American physician and vasectomy fanatic at the Indiana Reformatory, called "The Sterilization of Degenerates":

> Most of the insane, the epileptic, the imbecile, the idiotic, the sexual perverts; many of the confirmed inebriates, prostitutes, tramps and criminals, as well as the habitual paupers found in our county poor asylums; also many of the children in our orphan homes belong to the class known as degenerates.

Sharp asserts that the majority of cases of "insanity" have ancestral roots—again a Galtonian interpretation of heredity—and that marriage sanctions are not enough to prevent sex or reproduction. Sharp advocates vasectomy—of which he had performed 236 since 1899, taking three minutes without local or general anesthetic, and without any "unfavorable symptom." He provides quotations from eminent doctors to extend an equivalent sterilization practice to women and concludes that this "should at least give courage to others that are interested in the purity of the race."

I suppose there is something of an irony buried in Churchill's enthusiasm for purging alcoholics and the mentally ill from the population. He was a proud and hard drinker, quite possibly an alcoholic, throughout his life. He mocked those who criticized his drinking, including King George V, who announced he would abstain in support of the British troops in the First World War, and openly admitted he relied on alcohol. Furthermore, Churchill was plagued by periods of melancholy and mood swings. Posthumous diagnosis of profound mental ill health is always a minefield, but some have claimed that he suffered clinical depression, even bipolar disorder (though I think this is a stretch). Nevertheless, Churchill's behaviors may well have put him in his own category for enforced sterilization and absolute removal of freedom, had he not been born into hereditary power and privilege.

Churchill asked the Home Office to look at the sterilization law passed in Indiana in 1907 to consider how it could be implemented successfully abroad. The Feeble-Minded Control Bill was presented to the House of Commons on May 17, 1912. It proposed to implement the findings of the 1908 royal commission report, including the segregation of tens if not hundreds of thousands of people from society into asylums, and the criminalization of attempts to marry anyone designated as feeble-minded or one of the other criteria of unfitness.

I have labored the point that support for eugenics was broad, but that does not mean it was uncontested. Though wildly popular across political divides, eugenics was by no means universally supported, and this bill would not survive its critics.

Plenty of people vocally and publicly opposed the principles and the enactment of eugenics policies in the United Kingdom and abroad.

Just as some of the classic literature of this age references eugenics, activists and writers had significant impact on hampering the progress of eugenics in society. Perhaps the most lasting of these works is H. G. Wells's science fiction classic *The Time Machine*. Written in 1895, it's not really a story about time travel at all, in the sense that it doesn't deal with the consequences of altering chronological cause and effect, as happens in later sci-fi time-travel movies such as *Back to the Future*, *La Jetée*, *Avengers: Endgame** or *Hot Tub Time Machine*. Wells's time traveler experiences a dystopian future in the year 802,701, where humankind has bifurcated into two distinct species: the Eloi, a surface-dwelling fruitarian people descended from the Victorian upper classes, who avoid the dark and moonless night for fear of encountering the Morlock, apelike subterranean grunting cave-people whose ancestors were the working classes. The Morlock are seen at first as an underclass effectively in service of the fey Eloi: the machinery of state that they rely upon is maintained by the surly Morlock, but as the story progresses it is clear that this peculiar symbiosis is collapsing. The Eloi are degenerating, and the Morlock growing independent.

---

* Itself a film with very specific Malthusian themes. The main character, Thanos, explicitly sees population growth with limited resources as the root cause of all problems in the entire universe, and successfully enacts a cull of 50 percent of all people as the cure. This purge is undone only via the deus ex machina of corrective time travel.

Wells, an avowed socialist, vacillated on the subject of eugenics, and wrote fiction and essays about its moral and scientific folly, even though he also sometimes expressed sympathy and support for its implementation. Nevertheless, the best science fiction always acts as a commentary on the issues of the present, and one interpretation of the themes in *The Time Machine* is as a cautionary tale of societies as they evolve away from a recognition of the value of every human life.

One of the most effective campaigners against eugenics in the United Kingdom was the writer and Christian apologist G. K. Chesterton. He regarded eugenics as a tremendous evil that ran counter to his deeply felt Catholicism. Over many years, Chesterton wittily and forcefully wrote and lectured about eugenics—"a thing no more to be bargained about than poisoning"—and especially its anti-Christian rubric. He believed that an inherent property in the principles of eugenics as proposed by its advocates was that it targeted not the weak—whether that was assessed by somewhat arbitrary physical, behavioral or mental abilities—but the poor.

> The Eugenist, for all I know, would regard the mere existence of Tiny Tim as a sufficient reason for massacring the whole family of Cratchit. . . . The poor are not a race or even a type. It is senseless to talk about breeding them; for they are not a breed. They are, in cold fact, what Dickens describes: "a dustbin of individual accidents," of damaged dignity, and often of damaged gentility. The class very largely consists of perfectly promising children, lost like

Oliver Twist, or crippled like Tiny Tim. It contains very valuable things, like most dustbins.

That quote was among Chesterton's many thoughts on the subject published in a 1922 book whose title could not have made his opinion any clearer: *Eugenics and Other Evils*. His is a prescient view, I think born not of an advanced or predictive sense of biology that was at that time undiscovered, but of a political belief that would ultimately be ratified by genetics. The eugenics of this era was predicated on the conviction that certain traits were innate, but in fact in time we would realize those are mediated more by nurture than nature. All traits are heritable to some degree or other, but that does not mean that they are primarily genetic. We inherit our environment from our parents, family and peers, so for many of the traits that animated the eugenicists, the prospect of breeding them out of families and populations was always doomed to failure. Criminality can run in families, but there is no gene for it. Alcoholism can run in families, and while there are genes that increase the risk of addiction, there is no gene for alcoholism. You can have every one of those risk factors, but never become an alcoholic if you never drink alcohol. Poverty runs in families, but there is no gene for being poor.

Christianity, a religion born out of poverty, was inconsistent in its position on eugenics, just as it had been in its position on slavery in the nineteenth century. Both eugenicists and their opponents claimed religious principles as their guide. As today, the simplistic notion that this was a battle between science and religion is false. Galton, a Quaker by birth, was frequently dismissive of religion: he used statistics in 1872 to show the inefficacy of prayer* and praised Darwin for uprooting the "nightmare" of Christianity (something that arguably neither Darwin nor his work ever truly did). Galton's antagonism to the religious celibacy of eminent Englishmen of the Middle Ages stalling the inevitable greatness of the nation set eugenics against the Church at its inception. Yet many Christians in the early twentieth century saw eugenics as a way to deliver their religious principles, and not in conflict with the teachings of the Church. The Reverend W. R. Inge, the dean of St. Paul's Cathedral, and a weekly columnist for the London *Evening Standard*, was one such prominent and influential voice. The Eugenics Education Society (later the Galton Institute) had courted religious leaders for support, and wished to embrace "religion, in so far as it strengthens and sanctifies the sense of Eugenic duty." Reverend Inge took up that mantle, writing in the first issue of the society's magazine, *The Eugenics Review*, that the "moralist and the biologist may have a somewhat different standard of values, but they want the same thing—to make men better." Like so many of the eugenicists, he expresses fear of the fecundity of what he calls "degenerates": "I cannot say that I am hopeful

---

* By comparing the longevity of commoners and royalty, on the grounds that *The Common Book of Prayer* asks parishioners to ask that God should "endue the king/queen plenteously with heavenly gifts, and grant him/her in health and wealth long to live." He concluded however that this daily prayer had no effect, and royalty were "literally the shortest lived of all who have the advantage of affluence."

about the near future. I am afraid that the urban proletariat may cripple our civilisation, as it destroyed that of ancient Rome."

He reinterprets Galton's criticism of great men who chose to not have children:

> If such a man lives and dies unmarried we do not think any the worse of him. It never occurs to us that, in spite of his valuable contributions to literature, science, or what not, he has perhaps neglected the chief duty which God and his country required of him. . . . [I]t is the moral imperative of the Christian to embrace the new eugenics to challenge the natural degeneracy of the lower classes. It is not Christian, it is only barbarous and medieval, to say that cure is right, and prevention wrong. Be patient, my scientific friends, with us clergy, for we are the natural custodians of various race-traditions . . . our common enemy must be met with modern weapons.

G. K. Chesterton's view was the polar opposite. He and other Christians saw the problems that came with poverty as things to be fixed rather than eradicated. His writing on eugenics consistently highlights the value and sanctity of human life. It was Chesterton's lobbying, particularly of the Liberal MP (and member of the extended Darwin–Wedgwood pedigree) Josiah Wedgwood, that proved effective in stalling the Feeble-Minded Control Bill. Wedgwood described it as a "monstrous violation" of human rights.

The timing here is delicate. The Feeble-Minded Control Bill

was rejected, but the government quickly followed it with the Mental Deficiency Bill, which Churchill helped to draft. The bill, which was brought before Parliament on June 10, 1912, attempted to clarify the terms of the "mentally deficient" such that they could be categorized, and segregated from society, but not using the specifics of IQ, as had been adopted in the United States. These are the definitions used in the bill:

(a) Idiots; that is to say, persons so deeply defective in mind from birth or from an early age as to be unable to guard themselves against common physical dangers;

(b) Imbeciles; that is to say, persons in whose case there exists from birth or from an early age mental defectiveness not amounting to idiocy, yet so pronounced that they are inca-pable of managing themselves or their affairs, or, in the case of children, of being taught to do so;

(c) Feeble-minded persons; that is to say, persons in whose case there exists from birth or from an early age mental defec-tiveness not amounting to imbecility, yet so pronounced that they require care, supervision, and control for their own protection or for the protection of others, or, in the case of children, that they by reason of such defectiveness appear to be permanently incapable of receiving proper benefit from the instruction in ordinary schools;

(d) Moral imbeciles; that is to say, persons who from an early age display some permanent mental defect coupled with strong vicious or criminal propensities on which punish-ment has had little or no deterrent effect.

The bill also proposed a Board of Control, an Orwellian-sounding governing body whose jurisdiction was anyone subject to the bill's definitions, and which would "be charged with the general superintendence of matters relating to the supervision, protection, and control of defectives."

A few weeks later, in July 1912, eight hundred delegates met at the First International Eugenics Congress to discuss ideas and policy. The meeting was dedicated to Galton, who had died in January of the previous year, aged eighty-eight. Ironically, his own genius and eminence would not be passed down the generations: he was married to Louisa Butler for forty-three years, but their matrimonial pair bonding was childless.

The meeting was organized by the Eugenics Education Society and the University of London. Politicians and scientists from around the world met at the Cecil Hotel on the Strand and discussed their ideas at one of the peaks in the popularity and influence of eugenics. Future prime minister Winston Churchill was there, as vice-president of the meeting, along with former prime minister Arthur Balfour, who declared at a keynote banquet speech that the grand challenge of eugenics was persuading "the ordinary man that the task science had set itself was one of the most difficult and complex it had ever undertaken." Eugenics, he said, "depended upon facts—which ought not to be difficult to verify." There, yet again, is that misplaced confidence of politicians when speaking of science. It turned out that the facts of heredity, especially the Galtonian staple of nature and nurture, in all our studies of human biology, was and is the hardest button to button.

By now though, Churchill had been promoted to the position of First Lord of the Admiralty, the political chief of the Royal Navy, and had other matters to deal with. The eugenics movement had lost one of its primary advocates for the key principle of involuntary sterilization.

The Mental Deficiency Act became law in August 1913, but without enforced sterilization included. Instead, it ensured that people deemed undesirable by their categorization of being idiots, feebleminded or moral imbeciles would be separated and isolated from society in institutions, under the auspices of the Board of Control for Lunacy and Mental Deficiency. Only three votes were cast against the bill. It stood as law until 1959.

That cross-bench parliamentary support is notable. It is appealing to think that the eugenic maintenance of existing power structures in Britain was solely the preserve of right-wing ideology, but this was not the reality. Though it is obviously the case that upper-class people like Churchill and Galton were keen supporters, eugenics also fell within some of the principles of prominent left-wing thinkers and politicians in the years that followed (they were also frequently born of inherited wealth and power). The reduction of poverty via coercion or compulsory sterilization was not anathema to the socialism of prewar Britain; it was part of it. In 1903, the playwright George Bernard Shaw, champion of socialist virtues, wrote that the "only fundamental and possible socialism is the socialisation of the selective breeding of man." Many of the founders of the influential left-wing think tank the Fabian

Society were eugenics fans, such as Beatrice and Sidney Webb.* The traditionally left-wing press, notably *The Guardian* (then the *Manchester Guardian*) and the *New Statesman*, were also supportive. An editorial in the *New Statesman* in 1931 spelled it out:

> The legitimate claims of eugenics are not inherently incompatible with the outlook of the collectivist movement. On the contrary, they would be expected to find their most intransigent opponents amongst those who cling to the individualistic views of parenthood and family economics.

Perhaps the most confounding of these supporters is William Beveridge, who fits the educational pattern of the eugenicists to a tee: top private school (Charterhouse), math and classics at Oxford. In 1942, Beveridge authored *Social Insurance and Allied Services*—aka the Beveridge Report—which outlined the architecture of the forthcoming British welfare state, and the foundations for the National Health Service. In 1906, well before he developed this beloved system of free socialized health care for all and an economic model designed to serve the poorest

---

* The philosopher Bertrand Russell is widely reported to have suggested the use of color-coded "procreation tickets" as a means of controlling the population for the eradication of weaklings. While Russell was in support of sterilization of mental defectives, the only reference close to him of this alarming proposal is in correspondence between his sister-in-law Mary Smith and her future husband, Bernard Berenson, in 1894.

members of society, he was quite open in echoing the antiliberty views of the most ardent eugenicists of the time:

> [T]hose men who through general defects are unable to fill such a whole place in industry are to be recognized as unemployable. They must become the acknowledged dependents of the State . . . but with complete and permanent loss of all citizen rights—including not only the franchise but civil freedom and fatherhood.

THE MODERN SYNTHESIS

These bills, the legislation and the political views about society and its ills that they reflect all seem illiberal, arbitrary and draconian to our modern ears. They are the opposite of personal freedoms. They were popular though not unopposed, and with the removal of compulsory sterilization from those bills in the years up to 1913, Britain had dodged state-legislated eugenics by a whisker, at a time when countries around the world, notably the United States, had embraced it warmly. That doesn't mean that eugenics died a death in its birth nation. Galton's scientific legacy was still flourishing.

The Arthur Balfour Professor of Genetics post was created at Cambridge University in 1912, named after the prime minister who had given the keynote after-dinner speech at the First International Eugenics Congress in London that year, and it still exists today. The first person to hold this chair was Reginald Punnett, geneticist and inventor of the Punnett square, which we continue to use to work out which of the genes of two parents will be dominant in their children—it's how we teach genetics to schoolchildren. Punnett had given a presentation at that eugenics conference in which he offered scientific caution that the number of traits for which we had detailed hereditary information was limited, but that feeblemindedness was one on which we could act right away. This argument was based on the grounds that it followed a pattern of "single Mendelian inheritance"—meaning that like pea wrinkliness or flower color, there was one gene that accounted for the trait, and it was passed from generation to generation in a particular pattern that could be recognized in a family tree. A similar pattern of recessive inheritance was presented for the combination of epilepsy and feeblemindedness: "two epileptic parents produce only defective offspring."

Punnett's view was identical to—and perhaps influenced by— that of Henry Goddard, the psychologist who had formulated IQ as the metric for feeblemindedness in the United States. In 1912, Goddard had published a very successful book, *The Kallikak Family: A Study in the Heredity of Feeble-Mindedness*, in which he analyzed the pedigree of one family that had split into two lineages, one fine and upstanding, the other criminal, diseased and delinquent. According to Goddard, the source of this bifurcation occurred when Martin Kallikak, a pseudonymous Revolutionary War hero, was en route home to his genteel Quaker wife, when he took a short detour to sleep with and impregnate a "feebleminded" but attractive barmaid, with whom he had no further contact. In his clinic, Goddard had been treating

Deborah Kallikak, the great-great-great-granddaughter of Martin, since she was eight. His notes describe her as a "high-grade
feebleminded person, the moron, the delinquent, the kind of
girl or woman that fills our reformatories." Goddard produced a
detailed and exhaustive pedigree of the Kallikaks, starting with
Martin, and traced a perfect pattern of Mendelian inheritance
for traits good and bad. His legitimate family was bounteous
and successful, whereas his bastard progeny produced generation upon generation of poverty, criminals and mentally disabled "defectives," eventually including Deborah. With this,
Goddard concluded that the feeblemindedness of the Kallikaks
was encoded in a gene, a single unit of defective inheritance
passed down from generation to generation, just like in Mendel's peas.

Family histories were popular mechanisms for promoting
eugenic ideas, and the Kallikak case study was influential. The
trouble with Goddard's bestselling book was that it was a fiction. It wasn't true scientifically, which we will come to in a
moment, but it also wasn't true historically or genealogically.
While it was the case that Martin Kallikak's extensive legitimate family was packed with Galtonian eminent achievers—
men of medicine, the law and the clergy—the good-looking yet
feebleminded barmaid never existed. Instead, the so-called son
of Martin Kallikak was in fact the unrelated John Wolverton
(1776–1861), the son of Gabriel Wolverton and his wife Cather-
ine Murray. John Wolverton was a landowner, with educated
children. And while there were poor farmers and people with
disabilities among his descendants, by the twentieth century

there were also teachers, pilots and a banker.* They were not the miscreants Goddard had imagined.

The scientific assumptions about the root cause of their problems were also incorrect. Set aside the utter vagueness of the diagnosis itself. Whatever the clinical meaning of feeblemindedness, there are effectively no learning disabilities or mental health disorders that are the result of a single gene defect—I shall examine this further in the second half of the book. Goddard had discounted any possible environmental reasons for the disabilities he described, instead attributing them all to the innately inherited—that is, to genetics. Some recent analyses of this work suggest that some of the disabled members of the pedigree showed clear signs of fetal alcohol spectrum disorder, a condition triggered by mothers drinking heavily during pregnancy. That, very obviously, is an environmental cue and not primarily modulated by genes. The best characterized of this suite of conditions is fetal alcohol syndrome, which includes behavioral problems, reduced cognitive abilities, and characteristic abnormal facial features—which are clearly visible in the photos of the Kallikak children in Goddard's book.† Furthermore,

---

* It should be noted that Goddard accepted some of the criticisms of his work, and a decade later agreed that his conclusions were invalid.

† The paleontologist Stephen Jay Gould asserted that the photos had been tampered with to make them look more sinister. While it is true that the photos had been modified for the book—as was typical for the time—I have not come across any convincing evidence that it made the children's faces more disfigured or clinically persuasive, nor that that it was a deliberate deception by Goddard.

other nongenetic causes could reasonably account for the generational problems in the Kallikak family, such as malnutrition, which is social and heritable, and can cause severe mental and physical health problems.

But for those following in Galton's footsteps, who misattributed complex behaviors to simple genetics, the Kallikaks were sharp evidence for the root cause of mental defectives. These types of family histories were just what the Eugenics Record Office under Charles Davenport were interested in— pedigrees that revealed patterns of inheritance, and therefore a fulcrum on which eugenics could act.

In the United Kingdom, the work on eugenics continued. The Eugenics Records Office had been bequeathed by Galton to the University of London in 1904, and evolved into the Galton Laboratory for National Eugenics as part of UCL in 1907. The scientists there were critical of Goddard and, more generally, of the ERO in Cold Spring Harbor—because the work was shoddy, not because of any moral qualms they had. Quite the opposite: three papers that were written by the London eugenicists specifically criticized the slipshod approach of the Americans with the warning that it would "cripple the progress of eugenics."

The scientists who populated the eugenics labs of Gower Street in central London were some of the greatest who have ever drawn breath. As well as Galton's legacy in giving the world eugenics, some of his statistical inventions are still in use today, though they were relatively basic. Galton was never officially at UCL, but Karl Pearson was, installed at Galton's bequest and behest as the first Galton Professor and director of the eugenics program. Pearson is rightly considered the father of modern statistics. Anyone who has dabbled with math and stats from high school onward has probably used the Pearson Correlation Coefficient, one of the standard ways of comparing two variables. If you want to know if writing about race correlates with receiving racist abuse, or wearing a face mask correlates with infection rates in a pandemic,* Pearson is your man.

Pearson was a polymath in a different way to Galton, his mentor and patron. They had similar educational backgrounds—top English public (i.e., private) schools, Cambridge to read mathematics—but then Pearson embraced a ridiculous diversity of academic positions and pursuits: in Heidelberg he studied physics and metaphysics; in Berlin physiology; he studied German history and literature, and was offered a position in Germanic history at King's College, Cambridge, but instead returned to London to study law. He eventually settled on math, with a professorship at UCL in 1884. Pearson's 1892 book, *The Grammar of Science*, contains nascent ideas about the interchangeable nature of matter and energy, dimensional space and relativity, all of which influenced a young Albert Einstein. Pearson's impact on the history of science is titanic.

Galton took Pearson under his wing, a kindred spirit in seeing the power of measurement and statistics in revealing the hidden patterns in numbers. Pearson's subsequent devotion to Galton is cringeworthy. He believed that Galton would be remembered over Darwin, rather than reviled by those who know his work and forgotten by those who don't, as he mostly is today. After Galton died, Pearson wrote an incredibly detailed

---

* Strong positive yes, and strong negative yes, respectively.

and mostly unreadable three-volume hagiography. Unlike his mentor though, he was a freethinker, and devoted to socialist ideals, to the extent that he refused both an OBE and later a knighthood as an expression of his opposition to the monarchy. Like Galton, however, he was a committed racist and eugenicist.

"We had been defeated," Pearson wrote in 1900 in reference to the Boer War, "I may even venture to say badly defeated, by a social organism far less highly developed and infinitely smaller than our own." He adopted the familiar refrain that science will provide the answer to this political issue: "What part from the natural history aspect does the national organization play in the universal struggle for existence? And, secondly, What has science to tell us of the best methods of fitting the nation for its task?"

The answer to Pearson's mind was, of course, eugenics. He sums up his racism, math and paranoia in 1925 in a statement that foreshadowed the language of the Nazis. Jewish immigrants, he predicted, "will develop into a parasitic race.... Taken *on the average*, and regarding both sexes, this alien Jewish population is somewhat inferior physically and mentally to the native population."*

---

* One technique developed by Pearson, called Principal Component Analysis, is used every day in contemporary population genetics, often to show how genomic data—the As and Ts and Cs and Gs in our DNA—cluster around groups of people in ways that allow us to see their ancestry and migratory histories. Ironically, this tool, if used and understood correctly, comprehensively demolishes the biological concept of race, on which scientific racism was once so confidently founded.

Pearson retired in 1933, and the man who replaced him as Galton Professor sits comfortably in the top tier of scientists of all time. Ronald Aylmer Fisher spent a decade at UCL, but even before then had established his credentials as a brilliant and creative mathematician-scientist.*

Born in north London in 1890, Fisher also followed that route of top private school (Harrow) and math at Cambridge. His legacy is peerless. Much of what is taught on standard statistics courses in all universities is from Fisher: significance tests, the T distribution, the F distribution, maximum likelihood and much, much more. He had already coined the term "variance" in 1918, which describes how far numbers spread out from the average of a data set, and invented the concept of ANOVA— the analysis of variance—a powerful tool for simultaneously comparing range within and between data sets. Though these might not be familiar to anyone who doesn't use voluminous data in their daily lives, for those who do—including anyone who works in most branches of biology, medicine, epidemiology, psychology—these are screwdrivers, wrenches and levels, the basic tools with which we understand data. ANOVA, for example, is used everywhere, but if you want to know how fundamental it is to the modern world, you could pick any one of a hundred papers published since the spring of 2020 and ANOVA will be there in the methods sections, crunching reams of Covid

---

* Fisher's time at UCL does not seem particularly happy, and he left without much fanfare, despite having nurtured an unrivaled scientific legacy there. He took the Arthur Balfour Professor of Genetics post at Cambridge next, remaining there until 1957, when he retired to Adelaide, Australia.

data, everything from the efficacy of vaccines in clinical trials to the spread of the disease in specific countries.

Fisher's work was primarily in the service of evolution. Galton and Pearson had realized in the 1890s that natural selection occurs in populations. It is the changing frequency of different versions of the same gene in individuals in a population that allows us to see Darwin's ideas in action—descent with modification, from generation to generation. Therefore, the best way to work out the rules of evolution is to simulate large populations in numbers and equations, and apply subtle or significant pressures to see what happens—these could be a female preference for a bigger horn, or a color change in a butterfly's mimicry, or a population cut in half by a volcano or another natural disaster. Mendel had given us the units of heritable information— the gene—a few years earlier. But it was the fusion of Darwin's ideas, with Mendel's genes, with Galton and Pearson's populations, that set the scene for the biggest revolution in biology since Darwin. We call it the "modern synthesis," because unlike physicists who give their abstract and impenetrable work cool titles like the big bang or black holes, biologists are just terrible at catchy names. It was a synthesis of the ideas that had come before, but which hadn't quite joined up, held hands and slotted into place to explain how evolution precisely works.

Most of this revolutionary work was done by Fisher, his colleague J. B. S. Haldane at UCL (more on him in a minute) and Sewall Wright (formerly a student of Charles Davenport at the University of Chicago) and their endeavors effectively founded the field of population genetics. Most of this is detailed in the first half of Fisher's classic 1930 book *The Genetical Theory of Natural Selection*, a technical but essential textbook that lays out the foundations of the modern synthesis.

Actually, that's not quite true. The first seven chapters of *The Genetical Theory of Natural Selection* lay out the foundations of the modern synthesis. As well as being a peerless scientist, Fisher was a committed eugenicist. The final five chapters comprise a perplexing but full-bodied treatise on eugenics as the means to fix the looming decline of British society. "The deductions respecting Man are strictly inseparable from the more general chapters," he writes in the foreword.

Like so many of his eugenic brethren, Fisher dives headfirst into the fall of Rome and other classical civilizations. He laments the high fecundity of the lower classes as the source of the rot, and, using the 1911 census, compares the end of the Roman era to contemporary Britain by describing an inverse relationship between fertility and "value to society." He suggests tax incentives for middle-class people to have more babies, allowances proportional to the income of the father, and the abolition of benefits for large families with many children. Though he never endorsed involuntary sterilization, Fisher served on the board of the Committee for Legalizing Eugenic Sterilization, which advocated the neutering of "feeble minded high-grade defectives." I am unaware of a book that is so scientifically important and influential in one half, and so jarringly polemical and frankly bizarre in the other.

In this odd fusion, Fisher had form. His first foray into this field was when he helped establish the Cambridge University

Eugenics Society as an undergraduate in 1911. He published his
first academic paper in *The Eugenics Review*, at the age of twenty-
five, on the very Darwinian idea of sexual selection—"the love-
notes of song-birds, the flower-scented glands of butterflies,
besides the wonderful development of plumage and ornaments
among birds of every description." This article contains the
roots of what we now call Fisherian runaway selection—that is,
the process over generations where males acquire those preposterous horns or songs or dances to compete with other males
and impress females. Nestled in some of his hypothesizing
about how such exaggerated traits emerge are slightly rambling
hints as to how he might apply this evolutionary theory to people, with qualities such as stinky breath or ruddy cheeks as clues
to health, and therefore sexual attractiveness in potential mates.
But ultimately, he concludes, character surpasses base aesthet-
ics in humans. "Morality ceases to be arbitrary and dogmatic,"
he opines, "but takes its place as a particular formulation of the
requirements of the Highest Man—of our ultimate judgments
of human value."

Fisher was a theoretical biologist, meaning that he did little
or no experimentation. This is not to be scoffed at—science is
many things, including both theory and experiment—and he
had the utmost respect for the lab work of the experimentalists:
"The fact is that nearly all my statistical work is based on biological material and much of it has been undertaken merely to
clear up difficulties in experimental technique." This respect for
a solid theoretical base was part of the British culture of eugenics, and was inherited from the founder. In 1909, Galton had

issued a call to arms for the theoreticians to act as a bulwark against unfounded eugenics policy—as was already happening in the United States: "It cannot be too emphatically repeated that a great deal of careful statistical work has yet to be accomplished before the science of eugenics can make large advances." Fisher answered that call, in formulating the rules of evolution and plastering them on to humans.

For all three of these scientific giants—Galton, Pearson and Fisher—eugenics is applied science. It is not politics peripheral to their research, nor is it in parallel. Their attempts to understand the mechanisms of evolution are—at least in part—in service of the political eugenics movement. The idea that science and politics are independent is demonstrably and doggedly false.

FISHER VERSUS HALDANE:
CLASH OF THE TITANS

The same can also be said for those scientists who countered the eugenics movement. Just as G. K. Chesterton had pushed back against the political eugenicists of the Edwardian period, scientists did not unanimously back the supposedly data-driven justification for eugenic control.

The most notable of the scientific opponents was J. B. S. Haldane, a scientist of at least equal stature to Ronald Fisher. Jack Haldane was yet another product of British educational and class elitism: Eton followed by math and classics at Oxford (eighty miles from Cambridge, though culturally indistinguishable). However, his aristocratic family, his extreme privilege and his prodigious mind led him down a radically different political path.

Perhaps it was the bullying he endured at Eton, or his experiences as a soldier in the Black Watch, a Scottish infantry regiment, in the First World War, or maybe it was just that unpredictable soup of nature and nurture, experience and genetics, that drew Haldane to radical left-wing views and rejection of the hierarchical status quo of British society. Ultimately, he is arguably no less problematic politically than Fisher—in Haldane's case, for maintaining admiration for Stalin and his intellectually corrupt scientific deputy Trofim Lysenko, whose ideological principles ignored data and led the Soviet Union into famine and ecological disaster that would last for decades, ruining a nation and killing millions.

Haldane was at UCL from 1933 until 1957 when he immigrated to India. There he eventually would convert to Hinduism, dress in traditional Indian clothes, and die an Indian citizen: "sixty years in socks is enough." During his time as a scientist and science communicator, Haldane dabbled in physiology, on the causes of malaria, and even wrote a paper about genetics while in the trenches of the western front in France. He formulated ideas on the origin of life, and described the principle of size and scaling in animals in a classic essay "On Being the Right Size" in which he destroys the sci-fi notion that oversized King Kongs, Godzillas or giant spiders could exist.

Haldane had become a socialist in the Great War, and continued on this path, openly supporting communism from the

1930s onward. He toured workingmen's clubs delivering public scientific lectures, and even showcased a bizarre and frankly chilling Soviet film concerning the sensory stimulation and resuscitation of beheaded dogs.*

Haldane and Fisher started as full professors within months of each other, and pursued many similar and related scientific questions, in addition to founding the field of population genetics and the modern synthesis. In the Galton Laboratory, we undergraduate students listened to our tutors telling us scurrilous tales about Fisher and Haldane, and how much they loathed each other. There is one credible report of Haldane ostentatiously walking out of a Fisher lecture, but in general this hatred is hard to verify. Haldane had assisted in Fisher's recruitment, and the evidence suggests that they were cordial contemporaries rather than bitter enemies.

What is clear though is that Haldane's views on eugenics were ultimately radically opposed to those of his colleague. Haldane wrote frequently about eugenics, earlier in his life broadly in favor, but his 1938 book *Heredity and Politics* is assiduously opposed. It's an unsung masterpiece in formulating an argument with data and scientific theory, and dissecting the politics inherent in the science he was addressing. He is forcefully against eugenics, scientifically and politically (though with some caveats, which we will come to), right from the first page:

---

* *Experiments in the Revival of Organisms* is available to view on the internet, but I would not recommend it to dog lovers or the faint-hearted.

[T]he stringent measures which have been taken in Germany, both in the expulsion of the Jews from many walks of life, and for the compulsory sterilization of many Germans, are said to be based on biological facts. . . .

I do not believe our present knowledge of human heredity justifies such steps.

Haldane's arguments are both scientific and emotional. He adopts a similar stance to that of Chesterton in arguing that the groups targeted are vaguely delineated and the categories therefore open to abuse. Attacking American sterilization legislation from 1922 (formulated by Harry Laughlin), he suggests that by its criteria the world would have been robbed of Beethoven (on account of his deafness), John Milton (blindness) and Jesus (on account of his being poor, homeless and a "tramp"). Haldane acknowledges the inequality built into our existence, stressing its economic foundations, rather than the assertion by some of the eugenicists that for example "the unemployed represent relatively weaker stocks," and that he who is born "without manly independence affords the chief burden on the public purse."*

Haldane positions Fisher's innate ranking of people (from the later chapters of *The Genetical Theory of Natural Selection*) as in

---

* Haldane is a master of the snide footnote. His comments about Beethoven, Milton and Jesus are in three separate footnotes; in fact, the words "Beethoven," "Milton" and "Jesus" *are* the whole footnotes. In reference to the criticism of those born "without manly independence," his footnote, here reproduced as a meta-footnote, he says: "In my own experience, the majority of new-born infants are devoid of this quality."

opposition to the Declaration of Independence, whose words he attributes to Thomas Jefferson. In recognizing that people in reality are not born equal, Haldane addresses the argument made by eugenicists for the sterilization of people who are not permanently institutionalized, and who are more suited to monotonous work:

> An acquaintance informed me that he preferred feeble-minded men to look after his pigs. . . .
>
> I am of the opinion that a man who can look after pigs or do any other steady work has a value to society, and that we have no right whatever to prevent him from reproducing his like.

Haldane's clarity and foresight on matters of human heredity are stunning. He details the pathway of a series of diseases that run through family pedigrees in clear patterns, linked as they are to single genes that are defective, such as the hemophilia that plagued the descendants of Queen Victoria.* He uses these examples to demonstrate how heredity works, especially for congenital disorders that fit the same model of heredity as Mendel's peas. But it must be said that he does suggest, albeit in passing, that eugenics may yet have some value in the eradication of some, though not all, of these diseases.

---

* The defective gene in question, probably *Factor IX*, is carried on the X chromosome, so affects men more than women, who cover for the defective gene with a second version on their other X. Women are carriers, and do not suffer the disease itself.

The focus on the inheritance patterns of these diseases is a crucial point. Not only did the eugenicists downplay the role of the environment in the traits they wished to eradicate, but they also attributed complex disorders to single genes, as with the conjured evidence of the Kallikak pedigree. It's an idea that persists in contemporary culture, one that is baked into the way we talk about heredity: "Scientists Discover the Gene for X," is a typical formulation of the headline when a gene is newly found to *associate* with certain behaviors or diseases. The headlines that boast the lie that we have discovered a "gene for" something don't occur so often these days, because we've found most human genes now, and the revelations aren't coming as thick and fast as they were in the gold rush days of the 1990s and early 2000s, when new genes were being found week after week.* In this formulation though, X can be anything from homosexuality and transsexuality to risk taking and fear, from schizophrenia to autism. This type of thinking is symptomatic of a popular fallacy about genetics—we might technically call it monogenic determinism— where complex traits, diseases or behaviors are reduced to a single genetic cause, which imposes a fate on the bearer, whether it's being gay, daring, permanently petrified, autistic or epileptic.

---

* "Scientists Discover the Gene for Cocaine Addiction"—*The Guardian* (November 11, 2008); "Scientists Discover 'Transsexual Gene' That Makes Men Feel Like a Woman"—*Daily Mail* (October 27, 2008); "Study: A Gene Predicts What Time of Day You Will Die"—*The Atlantic* (November 19, 2012); "Gene That Can Scare You Out of Your Mind: Scientists Have Discovered an 'Anxiety Gene' That Makes People More Fearful"—*Daily Mail* (July 19, 2002); "Scientists Find 'Gay Gene' That Can Help Predict Your Sexuality"—*Daily Mirror* (October 9, 2015).

It's a fallacy in three dimensions: complex traits rarely have single genetic causes, they always involve the nongenetic environment and genetics is probabilistic, not deterministic. This is a key reason that the eugenics project was always on precarious ground: the conditions under scrutiny, whether it was feeblemindedness or epilepsy or alcoholism, do have a genetic component to them—almost everything in human biology and psychology does—though they are never single genes, and those genetic causes are rarely if ever deterministic. In the second half of this book I will dig deeper into what this means for our current understanding of heredity, and the refreshed contemporary conversations about eugenics with that modern knowledge.

It may seem indulgent to focus this pathway through the history of a complex cultural idea and practice on three mostly forgotten, arguably esoteric books: *Hereditary Genius*, *The Genetical Theory of Natural Selection* and *Heredity and Politics*, by three long-dead men who are largely unknown outside the fields that they directly influenced. The first is a Victorian proto-scientific study, typical of its era, with all the prejudices of British upper classes generously ladled into it. The second is mostly a textbook, describing theories many of which are still in use today, but is largely technical and mathematical. The third, a popular science book, aimed at a general readership, though it's not exactly Stephen Hawking's *A Brief History of Time* or Carl Sagan's *Cosmos*.* However, the influence of these three books

---

* In fact, *Heredity and Politics* is considerably easier to understand than Stephen Hawking's blockbusting bestseller, and it's much funnier too.

cannot be overstated. In a few pages we will add the German biologist Alfred Ploetz to this triumvirate, and his independent foundations of eugenics and euthanasia that inspired the Nazis, and we will see the pathway from idea to movement to genocide. The political movement of eugenics was always entwined with the scientific models of population control that began with Galton, muddled and flawed though those ideas were. On passing the Galton Professorship to Fisher, Karl Pearson was explicit about his own aims, that the study of eugenics and heredity should precede action.

The science served the politics.

> My endeavour during the twenty-two years in which I have held the post of Galton Professor has been to prove in the first place that Eugenics can be developed as an academic study, and in the second place to make the conclusions drawn from that study a ground for social propagandism only when there are sound scientific reasons upon which to base our judgments and as a result our opinions as to moral conduct. Even at the present day there are far too many general impressions drawn from limited or too often wrongly interpreted experience, and far too many inadequately demonstrated and too lightly accepted theories for any nation to proceed hastily with unlimited Eugenic legislation. This statement, however, must never be taken as an excuse for indefinitely suspending all Eugenic teaching and every form of communal action in matters of sex.

Perhaps in that quote is Pearson's dig at the Americans, Davenport's sloppy analysis, which the London Galtonians thought would undermine the credibility of the global eugenics project. I can't work out if this is an admirable stance or not—to criticize something grotesque because it is not rigorous enough in its grotesqueness. What is beyond question though, in those first two Galton Professors, is that their scientific brilliance, and particularly their invention of the tools of modern statistics and evolutionary biology, should be the foundations of the social action that would follow.

It is in the scientific arguments presented over decades that politicians based their policies, either proposed or enacted. Haldane is late to this party, and writes with the knowledge of the Nazis' growing programs of sterilization and euthanasia of undesirables and persecution of Jews. But his work is also the fusion of science and politics, very much the obverse of Fisher, Pearson and Galton.

As with the Americans, the persistent theme of the fall of existing civilizations is a lodestone to which the eugenicists are all magnetically drawn. In the United States, replacement of populations is the dominant threat; in the United Kingdom, it is more tethered to the rise of an underclass. Immigration features in both.

These are expressions of an eternal psychological bias known as "declinism"—the fantasy that everything in the past was better, and everything today is getting worse. This view was part of the intellectual discourse of the time, and maybe of all times. Madison Grant had written about it in his bestselling 1916 book,

The Passing of the Great Race, and Stoddard in *The Rising Tide of Color: The Threat Against White World-Supremacy* in 1920— note the unambiguous subtitle: non-Whites were threatening the entitlement of the White race's global domination. In 1918, the German philosopher Oswald Spengler had published the first volume of a treatise called *The Decline of the West*, with the second part coming out in 1922. Together they were highly influential, though not uncontested, presenting an argument that classical cultures became decadent, lazy and, when battered by barbarians at the gates, ultimately ate themselves. Perhaps the most influential of the works that describe the collapse of much revered civilizations is Edward Gibbon's *The History of the Decline and Fall of the Roman Empire*, the vast six-volume history of a period in which the demography and power base of the largest part of Europe (and beyond) fundamentally shifted on its axis. It hardly needs saying that it's a complex and long history. Roman centralized governance and military might waned from the fourth century CE, and by the end of the fifth the Roman Empire had petered out of existence, at least by comparison to its glorious, expansive heyday.

When scientists play historian, the risks are great.* All too often, in their attempt to understand history they fail to apply the same scrutiny to the evidence of the past to how they approach their scientific data. A position one can extract from Gibbon's epic history, if one were so inclined, is very broadly that what followed the fall of Rome is a dark age, priest-ridden and supersti-

---

* Forgive me.

tious, when knowledge stagnated and culture languished. Recall how Galton scolded the Christian Church of medieval Britain for castrating the continued brilliance of its best minds with enforced celibacy. These intellectual glooms lifted only when the Renaissance and Enlightenment and the Age of Reason reignited the intellectual, cultural ideals of the greatest civilizations that ever existed, Rome and Greece. Fisher reverently cites Gibbon, and laments the fall of Rome with little more sophistication and detail than I have done in this couple of paragraphs. For these men of power, Rome was a beacon of reason and culture, but the torch they carried for it was little more than a fetish.

THE ROAD TO THE HOLOCAUST

The adoration of the classical world, and the constant lamentation for its demise, runs through the eugenics movement like a seam. Fisher cites it, Churchill cites it, Reverend Inge (Eton and Cambridge) and Galton both do too. Perhaps the high-society upper-class classical education of the top English private schools drummed into them a knowledge of Rome, and its decline, and a belief in societies in which there was an assumed natural order and, with it, a hierarchy of human value.*

Inevitably with the story of eugenics, we come to the Nazis.

---

* It is a fetishization that has died out in English schools, now that Latin and classics are no longer a requirement for an education. I did seven years of Latin at school, which I loved, and then went to the Galton Laboratory to study genetics. Adoration of studying the classics is not the same though as revering the structures of the classical civilizations.

Adolf Hitler did not have that same classical education, but also found ideological bedrock in the ancient world, which takes us right back to the beginning. In the Spartans he saw a people governed by unabashed racialist principles—a *Völkisch* state. In Hitler's *Zweites Buch*, a sequel to *Mein Kampf* but not published during his lifetime, he praises the Spartan infanticide, as reported by Plutarch:

> The exposure of the sick, weak, deformed children, in short, their destruction, was more decent and in truth a thousand times more humane than the wretched insanity of our day which preserves the most pathological subject, and indeed at any price, and yet takes the life of a hundred thousand healthy children in consequence of birth control or through abortions, in order subsequently to breed a race of degenerates burdened with illnesses.

Hitler judged the Spartan military dominance over invading hordes as proof of their racial superiority, and the success of their strict population control: it is the destiny of the Aryan nation to inherit the power of the Spartans. The rise of Hitler, the Third Reich and the Holocaust are perhaps the most studied parts of Western modern history, and I don't propose to inadequately go over that ground here. For this analysis, the question is how, in just a few decades, eugenics had escalated from theory to genocide. Racial hygiene—*Rassenhygiene*—was a central pillar of the politics of the Third Reich, and though they reach their dark conclusion in the acts of Hitler's despotism, its roots in Germany had developed decades earlier, tied closely to the eugenics ideas growing in nineteenth-century Britain and America. Ultimately *Rassenhygiene* would include both eugenic sterilization and what is euphemistically described as euthanasia, where first children and later adults were killed in order to purify the German stock and remove the unfit, undesirable and unwanted from the gene pool.

The term itself was coined by Alfred Ploetz, a player equal in significance in the development of German eugenics as Galton was to the British and Davenport to the American. Born in 1860 into a wealthy family, Ploetz was from an early age enamored with the works of Darwin and the German scientist Ernst Haeckel. He also developed great interest in socialist ideas, and a passion for the legendary tales of medieval Teutonic tribes. In his memoirs, Ploetz confesses to swearing an oath under a tree with friends in which they promised to devote their lives to elevating the German people back to their glory days.*

His interest in socialism took him to America to work on a collective in Iowa, but instead of a utopian cooperative dream, he was crushed by the reality of what he regarded simply as bad people.

The unity of such colonies, especially those offering a large amount of individual freedom, cannot be main-

* I have drawn heavily on the outstanding work of the historian Sheila Faith Weiss, which includes detailed translations and analyses of the development of German eugenics into the Third Reich. The quotes from Ploetz are her translations.

tained owing to the average [quality] of human material
at present. . . . I came to the conclusion that the plans we
wished to execute would be destroyed as a result of the
[low] quality of human beings.

Ploetz quickly returned to Germany, and began studying
medicine and psychiatry, and drawing up his thoughts on using
the new sciences of selection and heredity to fulfill his hallowed
oath. Haeckel was a keen supporter of Darwin's ideas of natural
selection and evolution, and did much to promote and develop
them in Germany in the second half of the nineteenth century.
But the two men differed in one key aspect of their understanding of evolution: Darwin had settled on the idea that humans
are all one species, evolved from a common ancestor, whereas
Haeckel asserted a view of strong scientific racism, that ten
races of human had evolved in parallel, and with a clear taxo-
nomic hierarchy—White Europeans were clearly superior to all
and Africans were at the bottom, most closely related to apes.
This view was not uncommon at that time but had begun to
wane from its broader acceptance in the previous century. And
Darwin was right. Humans are one species, and do have a single
origin, in Africa.

However, Haeckel's particular form of scientific racism was
instrumental in cementing a worldview of the superiority of
White Europeans and, more specifically within that population, the so-called Nordic people. Ploetz assiduously subscribed
to this, as did many, and saw the Darwinian idea of selection as
a mechanism for improvement. He outlined his eugenics prin-

ciples in an 1895 book called *The Excellence of Our Race and the Protection of the Weak*, where he first uses the term "racial hygiene"—just as personal hygiene protected the individual, *Rassenhygiene* would protect a people.

It is here that Ploetz tries to apply the science of evolution to the idea of population control, just as Galton had done, specifically for the betterment of the Nordic—or Aryan—people. It almost shouldn't need saying, but there isn't such a population category as Nordic or Aryan. The various classification strategies that scientists and philosophers invented in the eighteenth century and later ended up bearing little relevance to how we now understand ancestry, populations and evolutionary history. Then, Nordic was a subset of Caucasian, and Ploetz and others asserted hierarchies even within the Nordic people. However, Caucasian is scientifically meaningless, and categories within Caucasian therefore also have zero biological utility. They can't be selected or eugenically purified, because they are invented. Tribes and lineages and families have cultural and social importance to many—today the commercial ancestry market is worth billions and relies on a weak supposition that the composition of your DNA will reveal the identities of your forebears in time and space. At best it's a fudge, a spell to bewitch your romantic and sentimental urges—to belong to a tribe of Vikings, Anglo-Saxons or other noble warriors. But really it's just gassy bullshit. What modern genetics has shown unequivocally is that while there are differences among people around the world, which manifest broadly as reflections of geographical clusters, the real story of humankind is one of continuous mixing, constant migration and shared ancestry. Our genomes can reveal much. It's in those letters of DNA that we can trace patterns of migration and sex, of families and conquest. But what we find more than anything else is wonderful impurity. The concept of race turned out to be biologically meaningless, and the idea of racial purity turned out to be a fantasy.*

As with much of the discourse in Britain about population improvement, the focus in pre-Nazi Germany was as much about class as it was about race (as we understand it today). Racial hygiene was initially concerned with declining birth rates, and the unwanted presence of people with disabilities who placed a burden on society and challenged Aryan dominance. Even within the culturally normalized support for eugenics, not everyone was an Aryan supremacist though, certainly not in the way that would come to dominate the ultimate manifestation of population control in Germany forty years later. An interesting side note is that Ploetz and other eugenicists regarded the great intellectual successes of Jewish people to be admirable, and wished for the marriage of Jews with Aryan Germans: "All anti-Semitism is a pointless pursuit—a pursuit whose sup-

---

* The modern conceptions of race were largely constructed in the eighteenth century, and asserted as biological categories. This has turned out to be incorrect, and modern genetics has shown conclusively that while there is plenty of variation among people and populations, these differences are not usefully reflected in our DNA. Furthermore, genetic genealogy has demonstrated that the human family tree rapidly collapses in on itself only a few centuries into the past, meaning that all humans have recent shared ancestry. This is detailed in my previous book, *How to Argue with a Racist* (2020).

port will slowly recede with the tide of scientific knowledge and humane democracy." In time, his views about Jews significantly changed, and his prediction was genocidally wrong.

Similar demographic changes were happening in Germany and Britain. The German empire was unified in 1871. With the industrialization of the Second Reich came growing cities, urbanization and an educated middle class—*Bildungsbürgertum*—fearful of a more visible, socially problematic people: mental defectives, alcoholics, sex workers. The German scientific and medical establishment had enjoyed a powerhouse century,* and had moved significantly into the domain of public health, especially with regards to communicable diseases. As such they saw themselves as custodians of heredity. Ploetz's initial version of this applied science was to imagine that the best germ cells—sperm and eggs—could be selected so that decline or degeneration could be averted before fertilization. In this way, the societal cost of poor-quality basic biology—genes—would never be realized in the treatment of defective babies, children and

---

* Titans of biology such as August Weismann, whose germ plasm theory had solved Darwin's mechanism of inheritance and set Galton and Pearson on the road to population genetics; Rudolf Virchow, to whom is attributed the discovery of cell theory—the idea that all life is made of cells, and cells can only be born from existing ones; Robert Remak, the Polish Jew who actually did that work, but was denied lab space and a professorship on account of his Jewishness. Virchow, though he weaselly claimed the idea for himself, honorably fought for public health matters, particularly on matters of sanitation. As a result of his social campaigning, the legend is that Otto von Bismarck challenged Virchow to a duel. Virchow chose sausages as the weapons of choice, one infected with roundworm larvae. Bismarck withdrew.

---

adults. A hundred years later, techniques predicted by Ploetz are used in clinics all around the world, but for the alleviation of suffering in individuals, both parents and children, rather than for the overall improvement of the people as decreed by the state. The impact of *The Excellence of Our Race and the Protection of the Weak* was not immediately huge, and at the turn of the century eugenics remained a relatively niche interest in Germany. But Ploetz persisted, and in 1904 launched the world's first eugenics scientific journal, and a year later the Gesellschaft für Rassenhygiene—the Society for Racial Hygiene. This was the first professional eugenics organization in the world, followed two years later in London by the Eugenics Education Society. The specific intention of Ploetz's club was to be an international model for how population sculpting could and should take place. In 1910, Ploetz announced his categories for eugenic regulation. As with all these endeavors in population control, the types of people of concern to eugenicists began in what we would now regard as a medically vague way, but rapidly expanded. Good families should have more children, explained Ploetz, and traditional motherhood encouraged; there should be the "establishment of a counterbalance to the protection of the weak by means of isolation, marriage restrictions, etc., designed to prevent the reproduction of the inferior"; transmissible (though as we now know, not genetic or heritable) diseases should be fought, as they poison the fitness of the "germ-plasm" (sperm and egg), especially syphilis, tuberculosis and alcoholism; there should be "protection against inferior immigrants and the settlement of fit population groups in those areas presently occupied by inferior elements," if necessary by force; and just to add to his unapologetic mix of science and policy, Ploetz backed the preservation of "military capabilities of the civilized nations."

In 1911, the ideas enshrined in Ploetz's eugenics society received a popular boost at the International Hygiene Exhibition, when five million people visited the Dresden expo to see the latest ideas in both personal and population hygiene. Ploetz's intention to go international was rewarded with follow-up shows in 1913 in Lima, Peru, and a year later in Genoa, Italy. In 1912, Ploetz was among friends in London at the International Eugenics Congress. There, among the British scientists and politicians—Pearson, Punnett, Balfour et al.—he presented a paper describing the threat from Slavs, with their strong Asiatic traits and high birth rates. He added that Poles, Hungarians and Russians were pressing westward, and the "preservation of the Nordic race is severely threatened as a result."

Eugenics in the United Kingdom and the United States enjoyed the endorsement of scientists as its popularity grew. In Germany the take-up was similarly keen, and it was the medical profession who signed up most enthusiastically. The reasons for this zeal are complex, but in the early twentieth century the status of doctors in Germany had waned, and many were unemployed and undervalued. With the rise of Nazism, an opportunity to rejuvenate the role of doctors in society and elevate their prestige was seized upon. The formation of the National-sozialistischer Deutscher Ärztebund—the National Socialist German Physicians' League—in 1929 was a step toward the unification of state and medical goals, and built upon the singular centralized organization of the eugenics movement that Ploetz had started.

In 1913, a third of the members of Alfred Ploetz's Society for Racial Hygiene were doctors or medical students, though membership was only 407 in total. By 1933, when Hitler seized control, it had more than trebled, to 1,300. By 1942, half of the doctors in Germany had joined the Nazi Party, more than 38,000 men and women who had sworn the primary principle of their profession: *primum non nocere*—first, do no harm.

It's hard to square those two ideas—a medical body committed to preserving life and not causing harm, and a political ambition that requires the sterilization or killing of babies, children and adults. Nevertheless, a key principle of German population control emerged in 1920 in a book by Karl Binding and Alfred Hoche called *Die Freigabe der Vernichtung Lebensunwerten Lebens* (a rough translation in English would be Permitting the Destruction of Lives Unworthy of Living). That term—*Lebensunwerten Lebens*—would develop over the next few years to be central to Nazi euthanasia and eugenics, and comprised the generically mentally ill, people with disabilities, alcoholics, homosexuals, interracial couples, criminals and people who were sources of "social turmoil," which would eventually include Jews, Slavs, Roma, Jehovah's Witnesses and other ethnic or religious groups.

In the years leading up to the chaos in which Hitler seized absolute power, the German eugenics movement grew and spread, legitimized by scientists, a rejuvenated medical establishment and invigorated by the eugenicists active in the United States. Other key players—scientists—picked up Ploetz's baton in those Weimar years, building a more sophisticated scientific outlook that focused on the threat from non-Nordic peoples with their high birth rates, and from within Germany those deemed unfit who would swamp the upstanding *Bildungsbürgertum*.

Of the other key players, Eugen Fischer is worth singling out. Together with Erwin Baur and Fritz Lenz, he wrote a 1927 textbook on racial hygiene called *Menschliche Erblichkeitslehre und Rassenhygiene* (*Principles of Human Heredity and Racial Hygiene*). Fischer was a prominent geneticist and eugenicist who had written previously on the mixed-race children of Dutch and southern Africans (described as Hottentots, who we can reasonably presume are from Khoisan populations) in German-colonized countries. He believed these hybrids were a threat to European culture, and so advocated a strict Aryan purity code. This work built upon his collaboration with Charles Davenport on mixed-race marriages a few years earlier. The connections between U.S. eugenics and German racial hygiene were already in place.

In 1927, the Kaiser Wilhelm Institute of Anthropology, Human Heredity, and Eugenics (KWIA) was founded in the Dahlem district of Berlin. Over the years until it was disbanded at the end of the war, much of the key scientific justification for the Nazis' ideology on racism, eugenics and racial hygiene was done here, with Eugen Fischer as its first director.

Though mostly funded by the German government, a significant chunk of money to pay for multiple research programs in the KWIA came from the Rockefeller Foundation in the United States. That foundation had been set up in 1913 by the oil magnate John D. Rockefeller, who was personally worth about 3 percent of America's GDP at that time (and may well be the richest American to have ever lived). Rockefeller had founded Standard Oil in 1870, just as oil and kerosene was becoming the fuel to which the United States would be utterly addicted for the next 150 years. At the peak of his wealth, Rockefeller controlled 90 percent of all oil production and distribution in the country.

The Rockefeller Foundation was established as one of the first large-scale philanthropic foundations, its core mission being "To promote the well-being of mankind throughout the world." An explicit remit within that mission was to offer financial support for science, public health and medicine. Rockefeller himself took little particular interest in the diverse and many projects that received grants from the charity's funds, which included campaigns to tackle malaria, hookworm and other infectious diseases, as well as education programs about birth control and sexual behavior. Although it might seem anomalous to us today, eugenics as it was perceived then fitted squarely within that remit of science, public health and medicine, and, as mentioned earlier, the foundation funded the Eugenics Record Office at Cold Spring Harbor, alongside regular contributions from the millionaire widow Mary Harriman and the Carnegie Institution.

The Rockefeller Foundation had global aspirations, however, and via its Paris office, had begun funding German scientists to the tune of many hundreds of thousands of dollars in multiple centers in Germany. In 1929, the KWIA was gifted money by the foundation to help fund the construction of its new building, and later to support it during the financial hardship of economic depression. Grants were also given to Fischer and to the German eugenicist Otmar Freiherr von Verschuer to work on twin studies, as per Galton's original observation—which Verschuer had reiterated in Alfred Ploetz's journal—that identical twins provided a natural experiment to extract nature from nurture. It is not clear what the motivation was for the Rockefeller Foundation to continue to fund German research into ideas of racial hygiene before and after Hitler's rise to power, and the charity had all but withdrawn support before the onset of war in 1939. Nevertheless, it had provided an essential financial base for research that was wedded to the ideology that defined the Third Reich's eugenics policies. Verschuer's assistant, Josef Mengele, would take this idea to heart, with unimaginably cruel experimentation on twins in the concentration camp at Auschwitz throughout the war.

The focus of Ploetz, and of Fischer, Lenz and Baur, was initially and primarily the preservation of the Nordic race; the rabid antisemitism that Hitler would embed in German policy did not appear to be a significant part of their worldview. However, it was clear that they saw the Nazi Party as the one that would embrace their eugenics dreams, and take them seriously. Therefore, they tolerated aspects of the policies of the Third Reich, such as its persecution of Jews, in exchange for the delivery of their more noble eugenic ideal.

Hitler was appointed chancellor of Germany at the end of

January 1933. Among his first acts was the passing of eugenics legislation. The Law for the Prevention of Offspring with Hereditary Diseases drew directly and heavily on the sterilization legal models written by Harry Laughlin, the zealot deputy of Charles Davenport at the ERO at Cold Spring Harbor. By this stage, eighteen American states had passed eugenics laws, so they were ahead of the game, and an inspiration to the German eugenicists. Laughlin was awarded an honorary degree from Heidelberg University in 1936.

The law also drew from Baur, Fischer and Lenz's *Principles of Human Heredity and Racial Hygiene*. Specifically, the process by which sterilization would be enacted included patients being referred to a eugenics court, made up of a health official, an expert in eugenics and hereditary diseases and a judge who was typically a Nazi Party member. Diseases and conditions suitable for sterilization included hereditary or congenital feeblemindedness, schizophrenia, bipolar disease, hereditary epilepsy, Huntington's disease, hereditary blindness, hereditary deafness, severe malformation and severe alcoholism. Under this legislation, which would be modified a couple of years later to ban appeals against the decision of the court, around 375,000 people had their sexual organs neutralized between 1933 and 1939. In September 1935, the Law for the Protection of German Blood and German Honor and the Reich Citizenship Law were both passed, collectively known as the Nuremberg Race Laws. Together they made it compulsory for partners to be screened for hereditary diseases, and encouraged people to check their spouse's family histories for diseases or insanity.

The Nazis were adept propagandists, and cinema was one method of normalizing the eugenic principles enshrined in the Law for the Prevention of Offspring with Hereditary Diseases. *Das Erbe* (*The Inheritance*) is one such film, a twelve-minute docudrama that played in theaters with feature films from 1935. It proclaims the very Darwinian ideas of the struggle for existence, as a kindly professor shows a series of nature films to a young blonde woman, presumably a student—stag beetles rutting, a dog killing a hare, a cat with a dead bird, two cocks sparring. The message is suddenly clear to the student: "Animals pursue their own racial policies!" she exclaims, and they laugh. Next, an animated family tree of none other than the Kallikak family, the fictional clan described in the bestselling book by American eugenicist Henry Goddard, and an explanation of the passing of the disease gene from generation to generation (recall that by this time, even Goddard himself had disavowed his own book). Then the film switches to shock images of disabled and disfigured people, denouncing the cost to society of such weakness, and a quote from the Führer himself: *"Wer koerperlich und geistig nicht gesund und wuerdig ist, darf sein Leid nicht im Koerper seines Kindes verewigen"*—"He who is physically and mentally unhealthy and unworthy, may not perpetuate his suffering in the body of his child."

Das Erbe's message is stupid and makes no sense scientifically. Dogs don't want hares to not exist: they just want to eat them. Birds do not impose a cost upon the lives of cats: cats just want to eat them. But the muddled intention of the film as propaganda is clear: nature purges the weak, and so must we.

A poster displayed prominently at the 1935 Wonders of Life expo in Berlin warns citizens of the fate of a nation if the weak are allowed to breed: an athletic blonde Aryan man shrinks over the years, and is replaced by a disabled and deformed man whose stature grows to a hideous giant, with the legend *So wird es kommen wenn Minderwertige 4 kinder und Höherwertige 2 kinder haben*—"This is what will happen when the high-minded have two children and the feebleminded have four."

As for Hitler himself, it is not precisely known what sources he read in developing his extreme version of eugenics, which would be enacted during the Third Reich. His views on religion are complex and subject to endless analysis, but we do know that he often expounded the superiority of science over religion as a creed, and ordained a version of social Darwinism and evolutionary thought at the core of Nazi ideology. For example, this is from one of his so-called table talk monologues during the war, in 1941: "Science cannot lie, for it's always striving, according to the momentary state of knowledge, to deduce what is true. When it makes a mistake, it does so in good faith. It's Christianity that's the liar."

We know that Hitler read *Principles of Human Heredity and Racial Hygiene* by Baur, Fischer and Lenz while serving time in prison after the Beer Hall Putsch of 1923, his failed coup d'état. Madison Grant's 1916 bestselling book *The Passing of the Great Race*—alluded to in *The Great Gatsby*, and described by Hitler as his "Bible"—was the first foreign-language book to be published in Germany after the Nazis came to power, with its crazed assertions about Nordic purity.* In 1937, Hitler took direct action against the impure, and ordered the compulsory sterilization of the *Rheinland-bastarde*. In 1919, just after the First World War had ended, French troops, some of African descent, occupied areas of the Rhineland. Some German women married these men, and others had babies with them out of wedlock. In *Mein Kampf*, Hitler asserts that these troops were sent by the Jews, and that their existence contaminated the Aryan nation. Robust numbers are not easy to come by, but it is estimated that up to eight hundred of these children, none older than eighteen, were arrested and sterilized.

Hitler's ideological commitment to the killing of the unfit or unwell as part of the purification of the Aryan nation goes back to at least 1933. In his trial in Nuremberg after the war, Karl Brandt, the Führer's personal physician, testified that Hitler believed the euthanasia program was a course of action that the public would not accept. Hitler told the National Socialist German Physicians' League that "[s]uch a problem could be more smoothly and easily carried out in war."

Hitler took a personal interest in ensuring that his wish was

---

* Hitler was also a big fan of Henry Ford, the profoundly antisemitic founder of the Ford Motor Company. Ford published anti-Jewish literature regularly in his newspaper, the *Dearborn Independent*, and promoted ideas from *The Protocols of the Elders of Zion*, a fabricated text from 1903 purporting to describe the secret plan for Jewish global domination. It was a fraud when it was created and one that remains part of replacement theory today. Ford published half a million copies for distribution in the United States, and Hitler praises him twice in *Mein Kampf*.

fulfilled. On February 20, 1939, a child was born to the farmers Richard and Lina Kretschmar, in a village near Leipzig. The boy, Gerhard, was severely disabled, with one arm, one or no legs (reports differ), he was blind and he suffered from convulsions. The parents, both Nazi supporters, asked the local Leipzig hospital to have the child killed. They described him as "a monster." The professor of paediatrics there, Werner Catel, informed them that mercy killing of this nature was illegal, but suggested that they petition the Führer directly. Hitler sent Brandt to investigate with the instruction that if the description of Gerhard's condition were accurate, then he was authorized to end the baby's life. Brandt and Catel concurred, and Gerhard Kretschmar was killed on July 25.

With this legal challenge passed, and with the direct authority of the Führer, the acts of euthanasia began within weeks. In October, this was formalized as a policy that became known after the war as Aktion T4 (named after the address Tiergartenstraße 4, where the Berlin offices that administered the program were based). Per Hitler's warning that public support could be certain only in wartime, the policy was backdated to September 1, the day the Nazis invaded Poland, two days before the French and British ultimatum to withdraw expired, and war was declared. Initially under Aktion T4, newborn babies and young children were thought to be the most appropriate starting point; the first rounds of murders included 5,000 children killed by exposure in unheated wards, or by poison gas, starvation or cyanide. Soon it extended to adults in six euthanasia centers, and people with disabilities, long-term illnesses, Jews and others were killed, legally. The numbers are not easy to verify, but an estimated 300,000 people were murdered under the policy.

THE AFTERMATH

The Nazis had embraced involuntary sterilization with laws derived from American legislation, and euthanasia directly from Hitler's absolute edict. In just a few decades, eugenics in Germany had escalated from theory to genocide. On July 23, 1944, Soviet troops liberated the Majdanek concentration camp. Auschwitz followed on January 27, 1945, also liberated by the Soviets. American troops entered Buchenwald on April 11 and Dachau on April 29, and the British freed Bergen-Belsen on April 15. Those soldiers were the first to see the true realization of the Nazi's policies, crimes of unfathomable cruelty and evil that were fueled, at least in part, by the ideas of eugenics. On May 8, 1945, Germany unconditionally surrendered to the Allies. The formal reckoning of the crimes of the Nazis began that November with the Nuremberg trials. These had been set in motion in 1943 at the Moscow Conference, when the Atrocities Declaration was signed by the foreign ministers of the Soviet Union, the United States and Great Britain, promising legal retribution in response to "evidence of atrocities, massacres and cold-blooded mass executions which are being perpetrated by Hitlerite forces."

It is worth noting that this document, which set out the criteria for prosecution of the Nazis, was largely drafted by

Churchill, who thirty years earlier had campaigned for British eugenics policies. The Nuremberg trials focused on the German high command, but a year later, a new court case began: *United States of America v. Karl Brandt, et al.* (aka the Doctors' Trial). Twenty-three doctors and scientists were charged with war crimes and crimes against humanity. The war-crimes charges included planning and performing the mass murder of prisoners of war and civilians of occupied countries, stigmatized as aged, insane, incurably ill, deformed and so on, by gas, lethal injections and diverse other means in nursing homes, hospitals and asylums, and participating in the mass murder of concentration camp inmates.

Seven were acquitted, seven hanged, including Brandt; nine were imprisoned for terms no less than a decade. One thing that emerged from the Doctors' Trial was what is now called the Nuremberg Code—a ten-point charter that outlines principles of research ethics when human experimentation is being performed. The code is not a legally binding document but is broadly supported by most countries as an important set of guidelines, primarily concerning the informed consent of participants in experiments, reduction of potential harm, and sound justification for the research. As it is not a detailed document, it doesn't specify the possible involvement of children or their parents in acquiring consent, but as the language stands, it is arguably the case that the gene editing of Lulu and Nana in China in 2018 violates one or more of these rules.

Josef Mengele escaped prosecution and died from drowning in 1979 in Brazil. Otmar Freiherr von Verschuer, Mengele's supervisor, was deemed a *Mitläufer*—a neutral Nazi fellow traveler—and was fined six hundred marks in his denazification hearing. It is possible that he had destroyed records that may have condemned him as a war criminal, certainly, he was a member of the Nazi Party, and his profoundly racist and antisemitic work from before the war remained in the public domain. He reinvented himself as a successful geneticist after the war, studying human heredity in Berlin, and later as director of the Institute for Human Genetics at the University of Münster. He died in 1969 having spent ten months in a coma following a car crash. Verschuer remained a member of the American Eugenics Society until his end.

In under forty years, an idea had grown from esoteric, academic and obscure to mainstream, popular and policy. The men who formulated these scientific ideas—Ploetz, Fischer, Lenz and many more—were not directly responsible for the Holocaust, the concentration camps, the murder of six million Jews and millions of others. Ultimately the policies of the shortest of the three Reichs were deranged and haphazardly drew from many sources, ideologies, cults and religions, and overwhelmingly from Hitler's commitment to Aryan superiority and fanatical antisemitism. However, Aryan supremacy and racial purity were the blood in the major artery pumping through the actions of the Nazis to the end, and eugenics flowed with them. The Final Solution to the Jewish Problem was an act taken way past the spurious scientific creed of eugenics—but it was nurtured by it.

This does not exonerate any of those eugenicists, German,

British or American. They acted within a culture of extreme
prejudice, classism and racism which they had contributed to
and validated, and continued to do so throughout the years
of murder and experimentation on humans in death camps.
Antisemitism may not have been a driving force for some of
the eugenicists, but their creation of scientific labels of the
unfit, of racial hygiene, of tiers of human value and of a scientific justification for the removal of people from reproduc-
ing their weaknesses in future generations, green-lighted what
would follow. *Lebensunwerten Lebens* directly gave rise to the
dehumanization of millions of people, including Jews, as *Unter-
mensch*—"subhumans"—and that enabled their persecution.
The Holocaust, the Final Solution, the concentration camps
were much more than simply eugenics enacted, but that scientific creed had fueled their existence. The pathway of eugenics
led directly to the gates of Auschwitz.

THE NEUTERING OF GALTON

The policies were deranged and genocidal and cloaked in a
science that was unproven. In the second half of the book, I'll
specifically address the consequences of mass sterilization and
Aktion T4 on the German people, as part of a broader examination of the question of whether their eugenics did or could
work. But before drawing this first half to a close, I wish to make
a short detour about the influence of several of these key eugenicists of the prewar era, and how we regard their legacies today.

Francis Galton's influence is unmatched in nurturing eugenics into the twentieth century, a nucleation site on which ideas began to crystallize. He's also been a major figure throughout my adult life. In the next few pages, I want to step out of the science and brief history and consider the presence of scientists and historical figures in our present. My undergraduate years studying evolutionary genetics in the 1990s occurred in the Galton Laboratory, where I (sometimes) listened to my tutors in the Galton Lecture Theatre; I also did my first research projects on the same corridor as a small glass case that held some of Galton's belongings—a head crank for craniometry, some notebooks with diagrams about the nose and facial shapes that constituted his data on female beauty.

The Galton lab had moved from its original home on Gower Street in central London by then, and was housed in a drab 1960s building on Stephenson Way just around the corner. It was all orange Formica fittings, institutional linoleum and frosted glass. The building is no longer there, and the collection of Galton's works and kit currently has no permanent home, though this is unconnected to the formal removal of Galton's name from UCL that occurred in 2020. Eighteen months before that, the provost of UCL launched an inquiry into the university's historical association with eugenics. Many academics and experts, myself included, provided testimony to the inquiry committee, who published their final report in February 2020.

There was minor drama around this event, which was intended to be a milestone in the history of the university: the report was penned by a minority of the committee and subsequently denounced by the majority, who refused to sign it because it contained historical errors and appeared to ignore some of the testimony given. The dissenters published their own report instead. This type of academic catfighting is not unheard of, but it is unedifying. Nevertheless, the recommendations that followed included the removal of Galton's name from buildings and academic positions at UCL. The Galton Professorship, at that time held by a scientist whose work is entirely unconnected to eugenics, was permanently retired, and the name of the first person to hold that position, Karl Pearson, was also erased from the campus. The Galton Lecture Theatre is now Theatre 115, the Pearson Lecture Theatre now G22, and the Pearson Building is now the North West Wing.

The genetics department, the direct descendant of the Eugenics Laboratory, subsequently and autonomously elected to change the name of its R.A. Fisher Centre to the UCL Centre for Computational Biology in 2020. Gonville and Caius College, Cambridge, where Fisher was an undergraduate, fellow and president, removed a stained-glass window commemorating his work.

Fisher's defenestration was not limited to the country of his birth: in June 2020, the U.S. Committee of Presidents of Statistical Societies permanently retired the R.A. Fisher Award and Lecture, and within days, the American Society for the Study of Evolution also announced that it would also be renaming the R.A. Fisher Prize.

These cultural shifts are contagious. The founding president of Stanford and keen eugenicist David Starr Jordan had a phalanx of things named after him too, most of which have been retired in the last couple of years. Indiana University (where he also served as president) removed his name from the biology department, a parking garage and a creek that flows through the campus. In February 2022, Jordan Avenue in Bloomington was renamed Eagleson Avenue after a prominent local African American family. In 2020, the Watson School of Biological Sciences at Cold Spring Harbor, Long Island, itself a cousin only a century earlier of the Eugenics Record Office, opted to rename as simply the School of Biological Sciences, after reconsidering James Watson's explicit and well-documented racism over many years.

These renamings or unnamings are never uncontroversial, and often are met with irate resistance. But the arguments that rage against these decisions are frequently not clever. The posthumous naming of a building, or the erection of a statue, is a political act in itself, and allows the greatness of historical figures to loom above us without context, and without any form of analysis or education about who those people were and what they did. In that sense, statues are the precise opposite of history. Did a single student who sat in the Galton Lecture Theatre or by the Jordan River ever stop to think who these men were, and why they apparently deserved buildings or creeks named after them, many decades after they had died?

One striking thing is how inconsistent these reappraisals are. The Galton Institute (of which I am a member)—formerly the Eugenics Education Society—evolved into the Adelphi Genetics Forum in 2022. In November 2021, Imperial College London considered (and ultimately rejected) a proposal to remove a

bust and the name of the nineteenth-century evolutionary scientist and vocal abolitionist Thomas Huxley from its campus,
on account of his views on racial hierarchies (views very widely
held at the time), while apparently not considering that the
whole university is named after the empire that oversaw slavery. Kellogg remains a global brand apparently untainted by
its weird sex-obsessed racist eugenicist founder, just as many
of us drive Ford cars despite the proud antisemitism of Henry
Ford, who is name-checked in *Mein Kampf* as a role model to
its author. The past is a dirty place, its protagonists are merely
people—evil, genius and everything else in between. We cannot
and should not abandon nor trash the scientific works of Galton, Fisher, Pearson, Jordan, Watson and the many others on
whose scientific shoulders we stand. Their techniques and discoveries are in constant use, for the betterment of science and,
by extension, all humans. These are formidable legacies. But the
time of carte blanche for the unquestioned heroes of history is
over. We can choose not to honor their names.

It is not without irony that part of Galton's financial endowment has been redistributed to fund research positions for fellows from minority and racialized groups who were targeted by
the eugenics policies of the past. I lecture at UCL about Galton,
Pearson, Fisher and the history of eugenics, and my salary is
derived from Galton's bequest. In a sense though, the precise
wording of his codicil, in which he outlines the purpose of his
endowment, is continuing to be fulfilled, deliberately, though
perhaps not as he intended. It states that his money should be
used to fund various activities, including to:

Collect materials bearing on Eugenics; Discuss such materials and draw conclusions; Extend the knowledge of Eugenics by all or any of the following means namely (a) Professional instruction (b) Occasional publications (c) Occasional public lectures (d) Experimental or observational work which may throw light on Eugenic problems.

My colleagues and I do all those things. There is an elegant irony in a man launching a field with specific intentions only to have that very same field flourish, reject and condemn all his presuppositions. Such is science.

Perhaps D. H. Lawrence's fantasy of a lethal chamber for the sick and the poor was youthful folly, for he did not return to those grotesque ideas in public. We cannot provide such excuses for Fisher. He truly believed in eugenics, and those ideals were persistent in his thoughts and work throughout his life. They are expressed in his early enthusiasms while in his twenties at Cambridge, in his field-defining work at UCL, and in his writings after the Second World War, when many had abandoned such convictions in the shadow of Hitler's evil. Fisher expressed sympathy toward the eugenics policies of the Nazis and defended the Nazi Otmar Verschuer. During the war, Verschuer employed Josef Mengele, and used samples obtained from Jews murdered in concentration camps. Genetics emerged out of eugenics labs, and some Nazis escaped convictions for war crimes by reinventing themselves as scientists, some in America, with impunity. Whether Fisher was fully aware of Verschuer's direct associations with Nazi experimentation on people is not known.

All these people and so many others of cultural and historical significance were great supporters of an idea we have learned
to despise. A common response to this truth is that they were
women and men "of their time." This is vapid. All people are of
their time, and it is impossible to be alive at any other time. It is
perfectly possible and indeed desirable to criticize the past, and
to criticize the views of people in the past through the lens of
our values and those of their contemporaries. That is the definition of history. Hitler was a man of his time, and was legitimately (albeit among political chaos) appointed to the position
of German chancellor in 1933.

Too often, the argument that the past was a foreign country where people did things differently, and that they were simply acting appropriately for that era, is deployed to end or avoid
discussion and debates, or to reinforce a cultural history that
serves only to make the powerful feel comfortable.

If you wish to understand the past and its legacies, then it is
not good enough to simply exonerate people's acts because times
were different. It's a specious attempt to gloss over difficult subjects. Crucially of course, the views and cultural norms of past
times were not universally held, just as they are not today. Not
all people held the same opinions, and racism, sexism and other
views that have waned through time were not necessarily universally supported. It is undeniable that many Germans were
dismayed by the politics of the Nazi Party and were opposed to
its racist, fascist regime. Significant parts of the German Catholic Church publicly denounced the Nazis' sterilization and
euthanasia programs. Similarly, the British imperialist Cecil

Rhodes had a despotic and unquenchable thirst for conquest, and this was viewed as contemptible by many. The *Manchester Guardian*'s obituary for him in 1902 was quite clear that he had "outlived the warmest of admiration that he thus won. For one thing his exclusive preoccupation with purely material considerations had led him terribly wrong, and through him and his press, had led this country terribly wrong too."

Even the great satirist and anti-imperialist Mark Twain reserved one of his sharpest insults for Rhodes: "I admire him, I frankly confess it; and when his time comes I shall buy a piece of the rope for a keepsake."

In years to come, our descendants will condemn us for what we believed, or for what we fought for or against, or for the things that we remained silent about. They will be right to, and we can forgive ourselves for our beliefs that will only become controversial or wrong as culture continues to change. But in our present, we can and must be honest about the giants on whose shoulders we stand.

A SCIENTIFIC CREED (PART ONE)

Josiah Wedgwood, the British politician who did more than perhaps anyone else to stand in the way of enforced sterilization in Britain, summed up the whole eugenics movement with a telling phrase: "legislation for the sake of a scientific creed which in ten years may be discredited." This single line holds the key to so many of the problems with eugenics, in his time and ours. It was a science project always intended for policy, a meeting of theory and public health, population control and political hegemony—science marshaled into ideology.

Why should we care about the obscure academic works of a few dead turn-of-the-century gentleman scientists? The woman on the street or the farmhand in rural Kansas surely did not care much about the writings of Charles Davenport or Ronald Fisher, just as a factory worker in Bavaria probably spent little time discussing the racial hygienism of Alfred Ploetz or Eugen Fischer. But these men provided a scientific justification for policy and invented a fuel to stoke its fires. This is how a niche idea became normalized. The cultural milieu provides the soil in which a political seed is planted. It is nurtured and fed by these men of science, who are revered and respected, and their ideas stabilize and become popular, and politicians latch on to them. Our societies elevated these men to the status of Masters of the Universe, who with data, funding and a new science could see truths beyond the day-to-day grind of life. Science in the service of belief.

Genetics is tough, and specialist, and requires years of training, hard graft and luck to succeed. And at the end of all that exploring, the least worst outcome we can hope for is that we are not wrong. We study the richest data set ever known—our DNA—using the most complicated structure in the known universe—our brains—all in the context of how people actually live. Lives are jumbled and chaotic, and the job of those of us who study humans is to fastidiously sift through this muddle on the hunt for answers and explanations of why we are the way we are. We must always expect science to be misrepresented, overstated and misunderstood, because it is complex, because the data is unending and because people are strange. What is galling is when science is misrepresented, overstated and mis-understood *by scientists*: not us average schmos who muddle along doing the best we can with our limited capabilities—but those who are the finest minds in their fields and, by extension, assumed to be among the smartest people who have ever lived. The whole purpose of science is to unshackle us from the biases that we are burdened with, and culturally bound to, so that we can see reality not as we perceive it, but as it really *is*. The eugen-ics project, conceived, birthed and nurtured by these brilliant minds, shows just how hard it is for us to cut loose from our prejudices, from politics, from morality and even hatred.

Science is, by definition, always only transitionally cor-rect. J. B. S. Haldane frequently volunteered his own ignorance in his writing, or at least his lack of confidence. This is a very scientific stance. Our results are always conditional, never abso-lute. The scientist who says "I don't know" is to be trusted much more than those who say that they do. Haldane is not saying, "I have no idea"; instead he is saying, "We do not have the data for this conclusion just yet, maybe we never will."

The most robust science survives the scrutiny of challenge, of falsification and of testing. But even when it does, as in the case of Darwinian evolution by natural selection, it is subject to ever-lasting refinement. Though I argue that science is always politi-cal, no matter how noble the principle that it is not, this very fact makes some aspects of science unsuitable as an immutable basis of policy. Or at least they are uneasy bedfellows. Politicians took succor for their ideologies from scientific confidence of the eugenics project. But they were not born of fact. The science was wrong, or at least not right enough to warrant the commitment to courses of action whose ultimate repercussions would be the absolute and irreversible infringement of personal freedoms, imposed by the state under the camouflage of the greater good.

I am not suggesting that policy should not be evidence based, but Galton's version of eugenics, and what followed it, was always flawed, a castle built on sand. Humans are nature and nurture. Nordic purity was a fantasy. Disabled people are not a threat to civilization. The eugenic elimination of diseases that are not heritable was scientifically baseless. Nevertheless, those lies bewitched the culture of that extraordinarily tumultuous and prejudiced age, as every age is.

The popularity of eugenics started to wane in the 1930s, perhaps when people began to see the real-world implementation of a bogus theory, in a global war that saw the persecution of millions, the slaughter of millions more, and the crushing of whole countries. In the United Kingdom, the Eugenics Education Society lobbied Parliament to amend the Mental Deficiency Act of 1913, which had failed to introduce compulsory sterilization. But the politicians rejected it, and in 1934 filed a governmental report that criticised the sterilization programs then in full swing in the United States for being genetically suspect. The English birth-control advocate Marie Stopes, once an ardent eugenicist who campaigned for compulsory sterilization, wrote to a friend that her love affair with this idea was over: "The word has been so tarnished by some people that they are not going to get my name tacked onto it." Haldane's *Heredity and Politics* in 1938 excoriated U.S. sterilization legislation as a "crude Americanism like the complete prohibition of alcoholic beverage."

In the United States in the 1930s, compulsory sterilization rates peaked, and would continue for many years after the war. But studies began creeping out that chipped away at the unshakable confidence of the eugenics cheerleaders, studies that showed that the feebleminded were born in equal proportions regardless of their social status, and that they did not tend to have many children anyway. By 1939, the Rockefeller Foundation had pulled the financial plug on the ERO, and America had abandoned the epicenter of eugenic thinking.

Nazi Germany was a fascist dictatorship, but in Britain and particularly in America, personal freedom is a fetishized absolute. Yet there was no liberty for the tens of thousands of people incarcerated in asylums. The pursuit of happiness was pollarded for the hundreds of thousands with the unwanted neutralization of their sexual organs, and permanent removal of their reproductive rights. Control was always at the heart of these politics, attempts to exert control on the zone where people, society and biology are knotted.

Eugenics, a scientific creed only a few decades old, had enjoyed a glorious global blossoming at the hands of the powerful, until the deranged murderous horrors of the Nazis were revealed to the world. And then, it was all over.

PART TWO

SAME AS IT EVER WAS

E xcept that it wasn't over at all.

The word fell from grace and would eventually become irredeemably toxic. But the idea persisted—and persists.

The great evil of the Third Reich was vanquished, and soon after the war had ended, the Doctors' Trial at Nuremberg would be the first legal reckoning of the euthanasia, eugenics and experimental programs enacted by the Nazis. Twenty-three people were tried, sixteen found guilty, seven hanged. Much of the trial was focused not on eugenics exclusively, but on the testing performed on people that had occurred in Germany and in the concentration camps. What the Doctors' Trial did successfully do was associate the principles of eugenics very clearly with the atrocities of the Nazis, which had the effect of sullying the word "eugenics," without necessarily calling out the practice itself.

Many keen eugenicists from Nazi Germany sailed into new positions with their histories barely an inconvenience. Some maintained their beliefs in eugenics as an unalloyed science of population betterment, but argued that the deranged policies of the Nazis had blemished the academic purity of their vision. Otmar Verschuer lived out his days in Germany, a eugenicist until the end. Fritz Lenz, former director of the Kaiser Wilhelm Institute of Anthropology and co-author of *Principles of Human Heredity and Racial Hygiene*, became a professor of genetics in Göttingen in 1946, and died there thirty years later, at the age of eighty-nine. Ernst Rüdin, the successor to Charles Davenport as the president of the International Federation of Eugenic Organizations, and the man who largely designed the 1933 sterilization law in Germany, was issued a five-hundred-mark fine. The enforced sterilizations in Germany were rolled into war crimes or genocide, but none of the actors in the design or enactment of U.S. eugenics programs would ever even be charged with a single misdemeanor.

Though the word fell from favor, abandonment of eugenics was not instantaneous either. In the United Kingdom, *The Annals of Eugenics* was launched by Karl Pearson in 1925, but only in 1954 did it mutate into *The Annals of Human Genetics*. *The Eugenics Review* evolved into the *Journal of Biosocial Science* in 1969, and its publisher the Eugenics Education Society became the Galton Institute only in 1989. The Swedish Institute for Racial Biology became the Department of Genetics at Uppsala University only in 1958.

Nevertheless, revelations about the depths of horrors of the Holocaust had a significant impact on the public mood, swinging it wildly away from the concept of eugenics, when it had been so popular just a few years earlier. But population control via involuntary sterilization continued in the United States, Canada, Sweden, Peru and many other countries, at the behest of governments, imposed upon the people whose reproductive autonomy was deemed undesirable. One might reasonably assume that the numbers petered out as eugenics legislation was eventually repealed. That is certainly true when comparing with the hundreds of thousands neutered in Germany under the Reich, and the millions murdered in the Holocaust. But the postwar numbers are still troubling, baffling and incomprehensible.

In America, states gradually abandoned the programs that had sterilized around 70,000 people since 1907—though this was still slow, and incomplete: Oregon enacted its final enforced sterilization in 1981, but California's enthusiasm for sterilization continued well into the twenty-first century. In June 2014, a state audit of sterilization procedures in female prisons concluded that:

> 39 inmates were sterilized following deficiencies in the informed consent process. For 27 of the 39 inmates, the physician performing the procedure or an alternate physician failed to sign the inmate's consent form certifying that the inmate appeared mentally competent and understood the lasting effects of the procedure.

The historical connection between eugenics and family planning persisted too. In 1970, President Richard Nixon signed the Population Research and Voluntary Family Planning Programs

Act, commonly known as Title X. Over the years, this policy has empowered community clinics and offered support in providing contraceptive services and counseling. But legitimate claims, such as by Marie Sanchez, the chief tribal judge on the Northern Cheyenne Reservation in 1977, said that sterilization, sometimes without their knowledge or understanding, had been performed on more than a quarter of Native American women of childbearing age. This state-sanctioned control of a population that had been historically persecuted by the ruling class looks indistinguishable from the worst examples of eugenics enacted in the previous hundred years. How is it fundamentally different from the actions of the Nazis? Why do these actions not qualify as attempted genocide?

The number of coerced sterilizations continues to fall in North America, but the wish to control the reproduction of those still deemed undesirable by some persists. In 2020, there were reports that up to twenty women had undergone involuntary sterilization in Immigration and Customs Enforcement detention centers.

In Saskatchewan, First Nations women were coercively sterilized as recently as 2018, acts which perpetuate earlier Canadian eugenics policies against Indigenous peoples. In 2021, news about the discoveries of mass graves of children at residential schools in Canada began to filter out into the world. Residential schools were an attempt by the state to erase First Nations people from the citizenship, by cultural assimilation. They operated between 1863 and 1998, and were primarily run by the Catholic Church, which elsewhere in the twentieth century had so vociferously

opposed eugenics. During that time, more than 150,000 Indigenous children were sent to these boarding schools. But they were
more like prisons, with unsanitary conditions, where children
were not allowed to speak their languages nor practice their customs, and were frequently abused. Thousands died. In June 2021,
the unmarked graves of 751 children were found at the site of the
Marieval Indian Residential School in Saskatchewan, which ran
from 1899 to 1997. In May, the bodies of 215 children, some as
young as three, were found near the city of Kamloops in British
Columbia, presumed to be pupils at the Kamloops Indian Residential School. These discoveries will continue.

Around the rest of the world, the imposition of control by
the state continues vigorously well into the present day. In
China, alongside their rules about how many children parents
are allowed to have, the Uighurs, one of the state-recognized
ethnic groups, have been persecuted for their religious and cultural practices, and it is estimated that hundreds of thousands
have been interned in so-called reeducation camps around the
country. A gynecologist in 2021 claimed that she had personally administered eighty sterilizations per day, five minutes
per woman, with the insertion of an intrauterine device. Other
reports claim that by 2019, the Xinjiang region "planned to
subject at least 80% of women of childbearing age in the rural
southern four minority prefectures to intrusive birth prevention surgeries (IUDs or sterilizations)."

In India, state-sponsored population control has been brutal. In 1975, Prime Minister Indira Gandhi declared a state of
emergency and President Fakhruddin Ali Ahmed issued a

constitutional edict of rule by decree. "The Emergency," as it is known, lasted for twenty-one months, in response to various internal disputes and turmoil, and wild population growth. Sanjay Gandhi, the prime minister's son, instituted a family planning initiative aimed primarily at men—vasectomies are easier to perform than tubal ligation on women—which purported to incentivize men to be sterilized in exchange for land or loans, or other deal sweeteners. Millions participated, but it is widely understood that coercion was the norm, and there is an overwhelming volume of credible reports of thousands of men being violently dragged away to undergo forced vasectomies. Estimates vary, but some reports indicate that six million men were sterilized in the first year of the Emergency, and many thousands died as a result of botched jobs. The policy switched to sterilization of women, perhaps on the grounds that there was a belief at that time that women were less likely to resist and protest their own bodily autonomy. Sterilization remains the primary form of contraception in India, and financial incentives are normal, as was once proposed by Francis Galton and Ronald Fisher. Their eugenics dreams came true. Most policies target poor women.

THE SPECTER OF EUGENICS

While we may like to say that eugenics fell out of favor after the Second World War, it is clear that the truth is much more complicated. Eugenics policies continue to be enacted around the world, as a means of population control, most notably in the two most populous countries on Earth. I've described just a handful of examples from dozens globally, but they clearly demonstrate that even in our enlightened world, where individual rights are cherished, reproductive control is frequently exerted unilaterally from the state to women and men, and our precious freedoms become specters.

My intention here is not to comprehensively record the continued incidents and development of involuntary sterilizations and other policies that in previous decades would qualify unequivocally as eugenics enacted. Instead, in this second part of the book, I want to ask different questions of where we are now. The named policy of eugenics may have waned after the war, but our knowledge of biology and heredity have increased, and we have invented precision tools to curtail the uncontrollable nature of reproduction at the molecular level. It should therefore be possible to ask whether eugenics could work today, and whether the policies of the Nazi and American eugenicists were effective. This line of inquiry leads to troubling new questions about what is possible in our unending wish to control reproduction. It asks us to consider whether a century of scientific progress has inadvertently enabled the dreams of the first eugenicists, and it forces us to look closely at the ways eugenics policies and ideas have reemerged in public discourse.

The eugenics labs and organizations of the West mutated into human genetics departments, and within these academic and scientific settings the rejection of eugenics as policy became universal. There is barely a single geneticist on Earth today who would claim to support eugenics as it was formulated in the prewar era. At UCL, we have a peculiar history with eugenics, given that so much of its development happened under our auspices. We teach this history, and have done for many decades. It is a core part of the curriculum for the hundreds of undergraduate students who take the basic genetics courses each September, and this has extended to be included in psychology and statistics modules in the last few years. These courses do not exist simply because the history is interesting. They are essential content because the work of the eugenicists lays the foundations for the science we teach and perform today. You can't understand population genetics without studying Fisher and Pearson, and knowing their views contextualizes the research they were doing. But it is not clear to me that this approach is widespread among the thousands of genetics departments that have sprung up around the world in the last few decades. There has been a resurgence in interest in the history of eugenics recently, which we welcome, and one of the purposes of this book is to spread the lessons of history and science beyond the walls of academic biology and history departments. Geneticists must know their own history, pernicious as it is, and so should the wider public: it is also the story of the twentieth century, and our present.

The emergence of human genetics from eugenics wasn't simply an exercise in rebranding or renaming. The fundamental science (in fact, pseudoscience) of eugenics research concerned heredity in humans, even if the motivation was political rather than either purely scientific or for the alleviation of personal pain. That is what many genetics departments became though—centers for the understanding of heredity, and with it the scientific bases for medical interventions. In attempting to pick apart the mechanisms of heredity, the patterns of inheritance, the designs of evolution and the fundamental ins and outs of sex, we were still working on the same questions that had been set up by the Victorian eugenicists. The difference was, and is, that necessary policy is not inherent to the science of human genetics, as it was for the ideology of eugenics.

And so, in working on the nuts and bolts of biology, we frequently focus on what happens when it goes wrong. What are the causes of inherited disease? What do they tell us about heredity? And what can we do to fix them?

That last question is inevitable. Science is the pursuit of truth, and that should be enough of a justification for doing any research. We do it because it's interesting, and if you don't agree you can fuck off, as the editor in chief of *New Scientist* once said. But that's not totally true. Science is in the service of knowledge *and* of people. The knowledge does not exist independently of people. There is, arguably, a moral obligation within the study of humans that the knowledge accrued should improve our lives, as societies and individually, and where possible, it should reduce suffering.

As we push at the boundaries of knowledge of genetics and inheritance and sex, it is impossible not to face up to that obligation. Diseases exist, and thousands have genetic causes, and we must treat, cure and, if possible, eradicate them. Any other position is morally indefensible.*

---

* Antinatalism is the opposing position, which argues that procreation is morally wrong because human life inherently includes suffering. However, abandoning procreation altogether seems like an unlikely course of action, regardless of any philosophical justification.

What happened in the genetics labs around the world is that we strove to understand the nature of diseases that may well have ensured the bearers were historically subject to eugenic enforcement. Those clumsy silos of feeblemindedness, imbeciles, morons or other general nonspecific disabilities that the eugenicists used were refined in the twentieth century as diagnostic precision became better, and more carefully and caringly deployed. But with those advances, there are new questions about how we can control reproduction, which come with serious moral valences that require informed societal discussion. And our burgeoning knowledge of genetics provides us with new data on how historical eugenics projects may have worked, or not. These areas are to be investigated thoroughly in making the argument that eugenics was not only a morally repugnant political ideology, but a scientifically phony one too.

In the next few pages, I'm going to walk us through some of the modern history of genetics. Some of this will be familiar to you, other parts not. Some of it has been known for a century, other parts were only discovered in 2021, the year this book was completed. Genetics is a vibrant living field of research, and all of it is relevant to the ideology of potential eugenic selection.

———————

Starting in the 1970s, the disorders that ran in Mendelian patterns in families became the first to be understood at a genetic level. Before the days of cheap and almost instantaneous DNA sequencing, geneticists studied family pedigrees—just as the Eugenics Record Office had done so clumsily before, but this time with a molecular precision. Back in the 1980s, it was a slow process; we would home in on a chromosome that seemed to track with the disease through the generations, and then gradually zoom in on that chromosome, inching toward a smaller and smaller region as if stalking an elusive prey, until it looked like there was a candidate gene in all that shambolic DNA (and our genomes are shambolic—evolution has no interest in neat filing, nor in aiding our exploration; its only concern is that genes function and can be passed on). Genes are written in DNA, but they are concealed, a precise lexicon chopped up with random letters that are written in exactly the same alphabet. Hunting for them in all that mess was—and is—a code-breaking challenge. The next step would be to see if that gene was also different in other people with the condition, and if it was, try to work out how it was spelling out a malfunctioning protein. You could test it in cells, animals or patients, and tease out clues as to why this one particular mutated gene, often as slight as a single spellimg mistake, was responsible for such suffering. This process took years of patient and meticulous work, and the invention of new tools and techniques at every stage. This is how we found that mutations in *dystrophin* cause Duchenne muscular dystrophy (DMD), that mutations in *FGFR3* cause achondroplasia—a form of dwarfism—and that mutations in *HTT* cause Huntington's disease.

In DMD, huge chunks of the *dystrophin* gene are just missing, meaning that the protein that normally helps hold muscle fibers together doesn't work. In boys with DMD, their muscles fail, eventually including those in the lungs and heart. In achondroplasia, bone growth is restricted because the protein coded for by the *FGFR3* gene is overactive, as a result of a single letter change, a G where there should be a C. In *HTT*, there is a short section of the gene where three letters of DNA are repeated—CAG—between eight and thirty-five times. Any more than thirty-five, and the result is Huntington's disease.

I type these words so casually, but they betray the immense volume of brilliant and innovative work by thousands of contemporary geneticists working at the very cutting edge of a new science, little more than thirty years old. I pick these three disorders because they were among the conditions that historically had been specified by eugenicists to be purged from society. But there are thousands of genes, and thousands of disorders. By the year 2000, we had characterized about a thousand genes out of an estimated ten thousand that cause single gene-inherited diseases. And we continue to do so, each time working out what the genetic fault is, how it fails or misfires in cells, and why that causes problems. With this new molecular precision comes a recognition that the faults in our DNA could be not only identified but fixed. Today, we have a solid, if wildly incomplete, grasp on thousands of diseases whose causes are individual genes. They came first because of their relative simplicity, compared with the complexities of other diseases, behaviors and traits. Nevertheless, progress there is accelerating, and some formerly controversial ideas about the nature of complex traits are no longer a scientific question at all.

THE NATURE VERSUS NURTURE DEBATE
IS OVER . . .

. . . And has been for decades. That other fundamental question of Victorian science remains one of the hardest we ask: What is the relationship between nature and nurture? The reliance on nature—that is genetic, as being the dominant determinant of so many desirable and unwanted characteristics—was central to the eugenics doctrine. Galton championed it over the environment, as did so many of his intellectual descendants.

Ultimately, that dogmatic stance played a significant part in the downfall of eugenics, at least in the United States. The Eugenics Record Office petered out of existence when its funding was cut in 1939, partly due to the retirement of its champion, Charles Davenport, and the failing health of his chief enforcer, Harry Laughlin, but also because of the growing recognition that their data was weak. It was not revealing what had been sought, which was clear, biologically hereditary patterns for the conditions under scrutiny. Many, if not most, were not governed by single genes—as had been confidently asserted for nebulous diagnoses such as feeblemindedness—nor were they exempt from profound social and cultural influence. Some geneticists of the time already suspected this, and that the eugenicists' models of inheritance were simplified to the point of being obtusely wrong. The American geneticist Thomas Hunt Morgan had spotted this early, writing in 1925:

> The pedigrees that have been published showing a long
> history of social misconduct, crime, alcoholism, debauch-

ery, and venereal diseases are open to the same criticism
from a genetic point of view; for it is obvious that these
groups of individuals have lived under demoralizing
social conditions that might swamp a family of average
persons. It is not surprising that, once begun from whatever cause, the effects may be to a large extent communi-
cated rather than inherited.*

We now know that nature and nurture were never in competition. When it comes to the complexities of human behavior, for
several decades most geneticists and psychologists have adhered
to what has been described as the first rule of behavioral genetics: "everything is heritable." All behaviors have a genetic component to them, and an environmental one too. What we really
want to know is how much of each, and how each influences the
other—how the script is performed.

The trouble is that this is a very hard question to answer, and
humans are a terrible organism to study. For moral reasons, we
do not experiment on humans in ways that could answer these
questions. Instead, we rely on observations, and try our best to
chip away at the question with ethical approval and without
causing any harm. One of the more useful techniques for flens-
ing nature from nurture is via twin studies. Identical twins are
born from the same fertilized egg cell, which divided and sepa-

---

* Nowadays we use the word "inherited" to refer not only to biological (that
is, genetic) inheritance. We inherit behaviors and our environment too, from
family and our social networks. Morgan here is using the word "communi-
cated" to convey nongenetic, social inheritance.

rated completely at a stage soon after conception. Thus two eggs are formed with identical genomes* (nonidentical twins are from separate eggs that grow in the same womb, and therefore have the same relatedness as siblings). In this way, the nature component of the nature–nurture axis is neutralized—in theory. Any differences between identical twins can reasonably be attributed not to genetics but to environmental influences.

This, by the way, was Francis Galton's idea. In 1874, he invented twin studies when he sent out a questionnaire to various hospitals around the United Kingdom in order to "collect data for estimating the respective shares that 'Nature' and 'Nurture' ordinarily contribute to the body and mind of adults, meaning by 'nature' everything that is inborn, and by 'nurture,' every influence subsequent to birth."

It was a brilliant innovation, typical of Galton.† He received ninety-four responses, and over the years since, twin studies have become a cornerstone for understanding human genetics and behavior. A further development came in the form of

* In fact, they are *near* identical, as various minor changes will accrue over development and life. These differences are not hugely significant for this particular discussion, but still emphasize the point that even when the starting place for two lives is identical genomes, they do not remain the same, and every human who has ever lived has a unique genome.

† Environment, or nurture, actually includes not just everything subsequent to birth, but everything that is not genetic, including all the influences from conception onward that are not directly overseen by the genome of the embryo. The other factor is randomness. Just stuff that happens that we can't really account for, such as the orientation of a baby in the womb, or an atypical birth.

identical twins who had been separated. Identical twins may have near identical genomes, but when they live together in a family, their environments are incredibly similar too. When not raised together, however, the social cues may well be more different than the genetic similarities, so similarities and differences in behaviors can be more readily partitioned into genes or environment. Twin studies have been a core part of genetics for decades now, and I don't wish to go into much detail for the purposes of this discussion. But it is worth noting that, powerful though they can be, they are not without significant problems. Because twins share the environments in which they are raised, just as brothers and sisters do, extracting the social from the genetic is not clear-cut. As for the relatively few separated twins projects over the years, many have been shown to be deeply methodologically flawed, or in a handful of cases possibly fraudulent.*

Separated twins are often raised in very similar socioeconomic environments, and always at the same time, so the environment may be less different than might be assumed. In some cases, the children were separated years after birth, so had devel-

---

* Notably in the work of influential IQ researcher and educationalist Cyril Burt, who over the years was frequently and plausibly accused by many of making up data but may have just been incredibly careless and *really* bad at his job. Either way, his work is now thoroughly discredited. Burt co-authored some of his studies with colleagues Margaret Howard and J. Conway, though there is a genuine possibility that either or both of them never existed. There is no record of either name at UCL, the given affiliation of their authorship. When challenged on this, he claimed that they both had emigrated and he'd lost their contact details. How about that.

oped together; in others, they lived near each other, were raised by different members of the same family, and even attended the same schools. Some separated twins had plenty of contact with each other as they were growing up, and others were later reunited and even lived together. In principle, separated twins studies are the perfect way of partitioning nature and nurture. But in the messy reality of lived lives, these studies are often deeply flawed. Hereditarians often lean heavily on twin studies to emphasize the genetic over the environmental, to swing the pendulum toward nature, and away from nurture.

Nevertheless, well-conducted twin studies remain a valid research tool, and historically served to reinforce the genetic component of heredity—we are not born blank slates. Instead, our canvas is already penciled from the moment sperm meets egg.

E PLURIBUS UNUM

By the early 2000s, it was becoming clear that single genes could not explain the wondrous sophistication of human beings, nor the cluttered mess of complex diseases. We began to develop techniques that could unearth the great maelstrom of genetic influence over complex traits. Finding individual genes had been arduous work, and so a new approach was adopted after the completion of the Human Genome Project (HGP). This grand scientific endeavor set out to read the entirety of a single person's DNA: genes, on-and-off switches, the repeats, the fossils, the junk, the repeats, all of it. It was a magnificent challenge, and came in under budget—about $3 billion, that is one dollar for every letter of DNA—in the summer of 2000 with the first draft sequence. Since then, it has been refined over and over again, until finally, in July 2021, a full annotated complete human reference genome was published, twenty-one years after President Bill Clinton's portentous announcement that we had finally created "the most wondrous map ever produced by humankind . . . the language in which God created life."

Mixed metaphors and clumsy religious allusions aside, the finer detail of our genetic makeup was now available to everyone. Galton would have loved it. It is the true metric of human similarities and difference, a colossal data set bulging with the most personal information possible, bearing the keys to unlock the patterns of inheritance and evolution, free for anyone to mine.

One of the many crucial technological leaps that came with the HGP was the ready availability of genetic data. We could sequence genes faster, cheaper and more accurately than ever before. Which meant more data, which meant more answers, which meant more questions. Soon after came the invention of a radical and revolutionary new way of digging around in DNA: genome-wide association studies (GWAS). Instead of hunting down a specific gene, you could instead take a group of people, the more the better, with the same trait or disease. By comparing their whole genomes, it is possible to look for bits within those twenty-three pairs of chromosomes that appear more similar to one another than to a control group who don't share the trait of interest.

Then you lay out the results on a graph, and see what stands out. We call them Manhattan plots, because they can resemble the New York skyline, with highs and lows, each building representing a location in our genome. The higher a peak, the more likely the DNA sitting at that location is specifically involved in the trait you're pursuing. If, for example, you did a GWAS for Huntington's disease, you'd get a large prominent spike in chromosome 4, because that is where the *HTT* gene dwells, and people with that dreadful disease all have a mutated *HTT* gene. For more complex diseases and traits, we find dozens or even hundreds of spikes, each indicating that the DNA in that location is probably relevant to the disease and trait. GWAS helped identify thousands of places in the human genome that are potentially important, that vary between people in populations, and that are associated with disease. With this innovation, we could unravel the genetic architecture that underlies every disease, not just the ones that have clear individual genes.

During the years of the Third Reich, schizophrenia was a specific diagnosis isolated for purging under the Nazis' euthanasia and eugenics programs, which began in 1933, and escalated with Aktion T4. The number of patients is not easy to verify, and the diagnoses are similarly opaque, but reasonable estimates from historians who have tried to account for the murderous chaos of Nazi Germany put the number of schizophrenia patients who were either sterilized or murdered at somewhere between 220,000 and 269,500. This was largely down to the work of Ernst Rüdin, who crafted the sterilization laws of the Reich but escaped prosecution with only minor inconvenience. In the 1930s, Rüdin worked alongside Franz Kallmann, a German Jew who fled Nazi Germany in 1935, and became a leading geneticist in the United States.* Together they developed twin studies with a focus on psychiatric illnesses, and just like Charles Davenport and so many others, concluded that schizophrenia was caused by a single gene.

They could not have been more wrong. Around 1 in 200 Americans today suffer with schizophrenia. It's a mental disorder characterized by episodes of psychosis, disorganized thinking, hallucinations and other deeply troubling symptoms. It's also highly heritable, meaning that the majority of the differences we see in risk for schizophrenia are genetic in origin—we know this from twin and family studies, which show that for example if one identical twin has schizophrenia then the other has a 40 percent chance of suffering it too. Indeed, the highest risk factor for schizophrenia is having a first-degree family member with it. So this means that there should be identifiable genetic differences in people with and without schizophrenia. The GWAS is a perfect method for finding these differences, and many studies with tens of thousands of schizophrenic patients have been performed over the last few years to track down the underlying genetics. Most recently, in 2018, one study found 145 individual DNA differences (that is, single letter changes—an A instead of a T, a G instead of a C, etc.) across the whole genome, meaning that there are at least 145 genetic variables that positively contribute to the probability of having schizophrenia.

---

* Kallman continued his work in schizophrenia in New York and maintained a belief that it was caused by a single recessive gene into the postwar era.

Unlike Rüdin and Kallman's assertion that schizophrenia was a monogenic disease, schizophrenia is in fact a hugely polygenic disorder—dozens of genes are involved, and none is causative. It is only in aggregation that these variants make up the increased risk for schizophrenia, which accounts for just a fraction of the proportion that is genetic and not environmental.

Regardless of this caveat, knowing that DNA is involved is an important step—thumbtacks planted in the map of the genome. When we discover these genetic variants, some of the questions that logically follow are: What is the function of that bit of DNA? Is it in a gene, or in a part of the genome that switches genes on or off? If so, what does that gene do? Why would the variation predispose someone ever so slightly toward having this terrible disease? These questions remain mostly unanswered, because we just don't know enough about human genetics and neuroscience yet. Or because we don't know how genotype and phenotype are related to each other. But as you might expect with a psychological disorder, many of those differences reside in genes involved in the brain—though it's worth noting that maybe *half of all* human genes are involved in some way in our big expansive brains.

Recall the first rule of behavioral genetics: "everything is heritable." Another one is the fourth: "A typical human behavioral trait is associated with very many genetic variants, each of which accounts for a very small percentage of the behavioral variability." Psychiatric conditions and psychological traits are polygenic, meaning that genes are involved, and they are many, but the different versions of genes play a cumulative but small role in the risk of the outcome. Schizophrenia showcases this well: the 145 bits of DNA found so far in schizophrenia patients represent a fraction of the variants that will be detected, for two reasons. First, these are only the common variants, the ones that can be detected in large groups of pooled patients; the more we look, the more rare variants we will find. But that is work still to be done, and for now, we can say with certainty that we have found only a small percentage of the genetic risk of schizophrenia. The second reason is that schizophrenia affects people all over the world, but different populations will reveal different variants for the same disease, and as yet our samples are somewhat skewed toward people of European descent—we simply haven't collected the genomes representing most people on Earth.

The proportion of heritability that can be identified is only a small percentage of the overall risk, which is also modulated by rare, still undiscovered genetic variants, and all of the social, cultural and environmental influences as well. It is perfectly possible to have every one of those 145 genetic variants, and never show the slightest sign of schizophrenia.

Alcoholism was another behavior that was targeted specifically for eugenic purification and it exemplifies this point too. Alcohol use disorder or alcohol dependency are contemporary and more precise diagnoses, and we know that they are heritable, because everything is. We even know that certain genes involved in the metabolism of alcohol are associated with addiction to it. In the latest studies (2018), GWAS identified eighteen different genetic risk factors for alcoholism, which accounts for a small proportion of the heritability, as per rule 4. But drinking alcohol is, of course, entirely socially mediated. You could have every single one of the genetic risk factors for alcoholism and never become an addict if you don't drink alcohol. The inverse is also true: you can have none of the risk factors and still become an alcoholic. These variants are specific to populations and environments. We might look for the same ones in different populations, but may well find that the phenotype of polygenic traits is not the same even though the genotype is.

Schizophrenia and alcoholism were both conditions specifically targeted by the Nazis. But let's go back further and consider the founding trait of eugenics, the one that Galton championed above all when taking his first baby steps into the field he founded: intelligence.

Few subjects cause as much ire and toxicity as discussions of intelligence and of its inheritance. The reasons for this are myriad, but they most certainly include the fact that intelligence—notably in the form of early IQ testing in the United States—was a significant criterion for eugenic intervention. Nevertheless, and I will be brief, there are certain things we can say about intelligence that should be uncontroversial.

The first is that it is measurable. Intelligence may be a difficult thing to define and, more broadly, cognitive abilities include many aspects of our behaviors, including reasoning skills, problem solving, abstract thought and learning capability. The IQ test, as originally conceived by Binet and Simon in France, and then translated and modified in America by Henry Goddard (the author of the 1912 Kallikak feeblemindedness folly), are nowadays standardized and designed to test reasoning, knowledge, mental processing speed and spatial awareness. The tests are culturally biased; this is well understood, and these days responsible psychologists try to account for that. As with all human traits, cognitive abilities are not evenly distributed. IQ across a group of people falls into a pattern known as a normal distribution, aka a bell curve, with the population average being set at 100 points, and around two-thirds of people being within 15 IQ points in either direction; 1 in 40 people is above 130 or below 70. There are other measures of cognitive ability, such as educational attainment—that is, how many years you stay in formal education. All the different metrics tend to correlate pretty well with one another.

It is a flawed system, but we do understand how it's flawed, and it's the best we have. IQ is a reflection of current abilities, and can change during life, including if you practice doing IQ tests. It is not immutable, nor is it absolute. It is a single metric for a complex range of abilities, but so is a drivers license, or a university degree classification. Like all behaviors, IQ is heritable. We are not blank slates for any of our behaviors, and intelligence is most certainly not exempt from that rule. One of the strengths of IQ as a metric is that it has been tested so many times over the last hundred years that we have masses of data on it, and when it comes to heritability, what we find is that about half of the variation we see in a population is down to genetic differences between people. That is not the same as saying that intelligence is half genetic and half environmental; it is that in any given population, there will be a range between the top scorers and the bottom, and we can sensibly attribute half of that difference to DNA, meaning that it is encoded in our differing genomes. This is scientifically uncontroversial.

And so, as with schizophrenia, heart disease, height and any other trait you might be interested in, nowadays we can find the actual bits of DNA that are influencing that difference. The genome-wide association studies for cognitive abilities started big, became huge, and are now gargantuan. The first major GWAS on cognitive abilities (in 2013, this time via the metric of educational attainment) featured 126,559 people and it uncovered 3 single-letter genetic changes of significance. Three years later, the sample size had doubled, but the genetic landmarks of interest had gone up to 74. Or there was the landmark 2018 study that had 269,867 participants and found genomic locations of note in 1,016 genes. Or the other 2018 landmark paper that had 300,486 individuals and found 148 genetic markers and 709 genes. Or maybe the big daddy, also in 2018, when the number was 1.1 million people and 1,271 places in the genome that were associated with cognitive abilities.

Finally, after a hundred years of searching, we had found the 709 genes associated with general intelligence. Or the 1,016 genes. Or whatever the correct number turns out to be. I am not disparaging this research; they are all terrific studies involving huge amounts of work, and powerfully demonstrate several things very clearly: (1) a lot of genes are involved in cognitive abilities; (2) we don't know what almost all of them do, but they do many things in many different tissues; (3) the differences we have found so far account for only a small amount of the variance in cognitive abilities, which means (4) we still have a lot of work to do if we really want to pin down the genetic architecture behind our minds.

Genome-wide associations are a fabulous tool for analyzing our DNA, but they are just a tool, and all tools can be misused. GWAS also provided much of the perpetuation of the myths of genetic determinism, where a single gene is misattributed as the cause of something messy and complex. The thousands of news reports I mentioned in Part 1, where headlines claimed, "Scientists have found the gene for . . . ," were almost always prompted by this type of study. Even that most august journal *Scientific American*, even in the era when polygenic traits were well-known and widely discussed, published a headline in 2016 " 'Schizophrenia Gene' Discovery Sheds Light on Possible Cause."

At least *Scientific American* had the good grace to put it in scare quotes. The A in GWAS stands for "association," meaning that the piece of DNA that sits at the bottom of a skyscraper in those Manhattan plots is *associated* with the trait. It doesn't say that it causes it, nor does it say how it works, what it does or even what that bit of the genome is doing.

Furthermore (and this is really getting buried in the haystack of human genetics), although we have sequenced millions of people's genomes, they do not represent a complete read. Instead, we look at the bits that are important and the variations that are common. These common variants are useful because they are informative when looking across populations. So, when we account for the influence of genetics for a trait in your sample, we're almost certainly looking only at the common variants, the ones we know many people have. For many traits and diseases, some, much or most of the real genetic variance has not yet been identified because it is harbored in rare genetic variants that we are yet to discover. Every human genome is absolutely unique, and statistically, that remains true for the entire history and future of humankind. The possibilities are endless.

This is why genomics is a growth field, and why the completion of the Human Genome Project was only the beginning of the era of the genome. This is why when identifying genes or gene variants that influence traits such as cognition, we're accounting for a small proportion of the genetic influence, each variant of which plays a minuscule role, which is associated with the trait, and measurable only at a population level.

The HGP was one of the greatest scientific endeavors of all time, a map that transformed the landscape of all biology and medicine. Our ability to read DNA and to interpret it were revolutionized in the era of the genome. We invented new techniques, for cheaper and faster analysis of genes and DNA, and reinvented them again and again. The first draft sequence, that of an unknown African American man, took many labs and about six years to compile. Just two decades later, anyone can have their genome read in a few hours for less than the price of a midrange flat-screen TV, and with that there has been a deluge of genetic information that swamps us on a daily basis, harboring more information than has ever been accessed in the history of the universe. One thing that we do know about human genetics with absolute confidence is how little we know.

CTRL C, CTRL V

We've spent the last few pages wading through some of the basics of genetics that we teach undergraduates today. Bear with me, as this upsum is the important bit for thinking about the questions of how eugenics could, would or did work.

There are two things happening in the twenty-first century that are profound and powerful and have become standard techniques in reproductive medicine and genetics. Both are relevant to the historical principles of eugenics. The first is the ability to identify the DNA that plays significant roles in traits and diseases. We know the genes that cause diseases and contribute to disease risk, and we can identify bits of our genome that are involved in complex traits. We know these mutations at a molecular level. There is no more precise way to understand the underlying cause of a genetic disease.

The second thing is that we have invented the ability to *change* DNA. In fact, nature has always had the ability to edit DNA, but in the last few decades we have co-opted and reinvented these evolved genetic proofreading and copy-and-paste systems for our own purposes. We've been editing, shuffling and remixing DNA in organisms* since the 1970s. Back then it was slow and clunky, and by the '90s it was faster but still inef-

---

* Initially in viruses, which are typically thought of as not alive since they do not have the ability to reproduce themselves without hijacking another living cell. I do not care much for this distinction, because no organism lives independently of another. Viruses have genetic code and use it in the same way that all living things do.

ficient and arduous. The years of the HGP improved our tools and helped standardize the techniques that by this stage thousands of geneticists were using every day. Extract some DNA from some tissue; cut it up into manageable pieces; copy it so you've got plenty to play around with; insert it into some bacterial cells; grow them and check that the bit you're playing with is the right bit; cut it up some more, and insert or delete bits of it. The analogy with copy and pasting in text documents is spot-on. With a click and drag of a mouse, I can replace a letter, word, sentence or paragraph simply with a bit of Ctrl C, Ctrl V.

The genetic versions of these edit tools were integral to the HGP and most molecular biology that followed. We got pretty good at gene editing, to the extent that entire genomes have been synthesized and rebuilt, with new bits, different bits or even fewer bits. Defective genes in one species have been corrected and replaced in another. Bacteria have been reengineered to produce drugs, food grown to be pest resistant with higher yields. And with the advent in recent years of a field broadly known by the contronym synthetic biology, genes have been commoditized and standardized to fit together like Lego. Anyone with basic lab equipment can piece together bits and bobs from multiple species to build a new living tool with a specific purpose— such as to detect pathogens in the environment, digest plastic waste, break down oil spills or produce clean carbon fuels that have been grown instead of dug out of the ground. Even if you don't realize it, genetic engineering is part of everyday life and never more so than today. By the time you read this, with luck, planning and political will, most readers will have received one of several possible vaccines for Covid-19. None of these essential treatments was developed or manufactured without gene editing as a basic and essential part of that production process.

———————

It is the second of those examples that is particularly relevant here: the ability to change—or even correct—genes. The concept of gene therapy has been around since the 1960s, but it wasn't until the first disease genes were characterized in the late '80s that it became a real prospect. The idea is that if you have a gene that is faulty and causes a disease, and we understand what is wrong in that gene, then perhaps we can introduce a corrected version and alleviate or even cure the symptoms. This is a medical intervention that treats the cause and not the symptoms.

But it turned out to be much harder than was initially thought, as everything does in biology. One of the biggest problems was delivery: How do you get the corrected gene into the tissue where it is most needed? Gene therapy is an ongoing project for conditions such as DMD or cystic fibrosis, and scientists and patients eagerly await progress. A very clear distinction should be made here between gene therapy as described above and potential gene editing that might interest the eugenicists from history. Treatments designed to repair faulty single gene diseases are called somatic gene therapies. The somatic cells are all of those that are not sperm or egg, which are known as germ cells. From an evolutionary point of view, our somatic cells are simply the husk that carries the germ cells, and only our germ cells contain the genetic information that will be transferred into future generations. Almost our entire existence serves this purpose. A person who had undergone somatic gene therapy would not transfer the corrected gene to their children. It is not transgenerational.

If, however, you were to correct the gene in an embryo, before all that amorphous blob of cells has begun to differentiate into our various body parts, then every cell would contain the corrected gene, including the cells that might become sperm or egg. Therefore, altering a gene in an embryo could result in a transgenerational genetic change.

Even gene editing has radically changed in the last decade, with an invention by two scientists, Emmanuelle Charpentier from the Max Planck Institute for Infection Biology in Berlin and Jennifer Doudna from the University of California, Berkeley. They came up with the new gene editing tool CRISPR-Cas9 (almost universally referred to and pronounced as "crisper"). It was derived and modified from a natural bacterial antiviral defense mechanism that targets the genetic code of an invading virus based on previous infections and incapacitates it—a form of bacterial acquired immunity. What Charpentier and Doudna (and others also involved in CRISPR's development) did is modify and commoditize this system as a means of gene editing in any organism. As long as you know the sequence of the bit of DNA you want to alter, you can send in a CRISPR guide, which will seek out exactly the piece of DNA you're after and modify, replace or delete it as required. I'm not a great one for breathless hype in science, but this technology is truly world-changing, and in less than ten years has already transformed gene editing across the board.

It's revolutionary for two reasons. The first is that it's really easy to do. A gene editing project that took weeks, months or even years a decade ago can now be done in a few days by someone with not very much molecular biology experience. Indeed, a major part of my own PhD project twenty years ago, which took the best part of a year to do—and failed—was repeated in the space of about three weeks last summer by a work experience student. So it goes.

The second reason is that it is precise. You can CRISPR a single letter of code, change a T to an A, and potentially correct a mutation that for all of history until this moment has caused untold suffering and, under the policies of various regimes through time, persecution.

BIRTH CONTROL

One of the other world-changing technologies that emerged out of former eugenics labs was the invention of prenatal screening. All pregnant women in developed countries are offered this as standard nowadays, with ultrasound scans to check the health of an embryo in its mother's womb, and more specifically to look for any telltale signs of physical developmental issues. The most well-known and readily available of these is the nuchal scan, a visual check for a thickened part of the neck in a fetus that is indicative of one of several chromosomal abnormalities. The most common of these is Down syndrome, which affects

about 1 in 800 live births, but it can also indicate a baby growing with Patau syndrome or Edwards syndrome (which are
both rarer, serious and predominantly lethal disorders). Or it
can indicate Turner syndrome, in which females lack a second
X chromosome (instead of being XX, they are X0); these women
have some physical indications and health complications, such
as infertility and heart problems, but are intellectually typical.

Our ability to look into the womb and into the genetics of
an unborn potential life took a huge leap forward with pre-
implantation genetic diagnoses, or PGD. This was invented in
genetics labs in the late 1980s, following on the heels of in vitro
fertilization (IVF)—so-called test-tube babies—and is now a
mainstay of reproductive medicine. The idea behind PGD is
this: a fertilized egg is one single cell, with a complete genome
compiled from the half genomes of sperm and egg. From that
point on, the cell will divide, and divide again and again, each
cell of that growing embryo also containing a complete genome.
Over the course of days, weeks and eventually nine months, not
only will those cells have divided innumerable times, but they
will have specialized to become all of our different bodily tissues, made up of millions of different types of cell. But at the
very early stages, the cells are totally unspecialized, and have
the potential to become anything, be it brain, bone, skin or
placenta—at that point, we call them pluri-potent. So, for a few
days after sperm and egg meet, the cells form what we call a
blastocyst, a ball without much indication of what is going to be
what, and even which end is going to be head and which is tail.

If you can get hold of that blastocyst, either by fertilization occurring in vitro, or by a process of uterine lavage, you can take a single cell (or a few) from that ball without damaging it—they called this a biopsy in the early years, and it is rather like any biopsy you might get for a questionable mole, or in a cervical smear test. You can then examine the DNA in the extracted cells and try to find out if the embryo is bearing any specific genetic diseases. This of course is not available to all, as that would be an expensive and totally unnecessary exertion of embryonic examination. PGD typically follows genetic counseling for potential parents who have known familial histories of inherited diseases.

These techniques were pioneered by many researchers all over the world, but the most significant advances were made by Leeanda Wilton at the University of Melbourne, and many truly groundbreaking scientists in the now hugely expanded genetics departments at UCL, including Joyce Harper, Audrey Magneton-Harris, Cathy Holding, Anne McLaren, Joy Delhanty and Marilyn Monk, all part of none other than the Galton Laboratory.

They were looking at single gene disorders in human embryos within a few days of conception, diseases such as Tay-Sachs and sickle cell anemia, and laboriously performing a then brand-new technique called PCR, which photocopies specific chunks of DNA so that we've got lots to play around with. You will know PCR as the tool that has tested millions of people for Covid-19 after a retch-inducing, eye-watering nasal swab, which gets processed in a desktop machine that takes less than an hour. But back then, they were doing it by hand, carefully pipetting and incubating tiny volumes of colorless liquids containing the entire genomes of lives as yet unlived. The principle of PGD is this: a couple with a history of one of these diseases could use IVF to generate several embryos, each with a risk of bearing the disease determined by the shuffle of genomes in the formation of sperm and egg. But by testing cells from each embryo using PGD, you could identify those not bearing the faulty gene, and implant only them. If those pregnancies were successful, the disease would be erased from that bloodline.

The first clinical application of PGD on humans was performed in 1988 by a team a few miles south of the Galton Laboratory, at London's Hammersmith Hospital. Embryos from two couples with familial histories of X-linked diseases (that is, they affect only boys) were tested by Elena Kontogianni, under the guidance of Robert Winston, and tested for the presence of a Y chromosome. Only females were selected for implantation, and thus both mothers bore girls, three of whom survived, and are now in their thirties.

The disorders they were avoiding are called X-linked mental retardation and adrenoleukodystrophy—a disease with a huge range of symptoms, including emotional instability and disruptive behavior in children, and descent into a vegetative state in adults and children as young as three. Though it is impossible to know precisely, these are almost certainly two conditions that would have been lumped into the silos of feeblemindedness in the age of eugenics. X-linked mental retardation accounts for about 16 percent of all cases of intellectual disability in males. These diseases ran in the families of those first PGD girls.

But with this medical intervention, both sets of parents had children free from diseases that fifty years earlier in various countries would have resulted in their being sterilized or, more likely, murdered by the state.

THE TOOLS OF GENETICS AND
REPRODUCTIVE MEDICINE

Let's briefly recap some of the technologies and interventions currently available to us. If a couple have histories of inherited disorders in their families, the first step would be genetic counseling. Experts in medical genetics look through the pedigrees of families with histories of specific diseases and try to work out the patterns of how those traits fall through the generations. In doing so, they can calculate the risk that each parent is carrying a disease gene, and therefore the risk that one of their children might have a specific disease. This was precisely the step that so much of the eugenics projects of the late nineteenth and early twentieth century relied on, but so many of their crucial pedigrees were just poor, and we are much better at this today.

The next stage could be pre-implantation diagnosis. An in vitro fertilization can be screened for sex and for an ever-increasing number of genetic diseases. Embryo selection comes next, where only those without any genetic abnormalities or disease genes are implanted into the mother's womb. The pregnancy can then be monitored for normal progress, including the prenatal screens for Down syndrome and other potential abnormalities.

I want to be explicitly clear here. In my considered opinion, none of these interventions are eugenics. What they are is medical techniques specifically conceived and designed for the alleviation of suffering in individuals. They are medical treatments that reduce the risk of serious illnesses, and give options to parents who wish to have children but, for no reason other than blind luck, carry a higher risk that their kids (and by extension their family) will suffer hardships or reduced mortality well beyond that of a typical life. These techniques were developed in the labs that evolved out of the eugenics projects, and in some broad sense carry historical resonance with the attempts of the eugenicists to control biology. They are also tools that we can presume would have been of great interest to the eugenicists, as they elicit this control over the causes of disease and have the potential to reduce the burden of those diseases to society. But the decision to terminate a pregnancy because of a prenatal diagnosis is something that I believe is an absolute personal choice and should be an unstigmatized right for women and parents. To do so is not eugenics.

In parallel, we have invented genetic editing tools with unprecedented accuracy, capable of altering single letters of DNA, in any organism, including humans. The legality of gene editing in humans is strict and almost entirely forbidden. To do experimental genetic engineering work on human embryos is incredibly tightly regulated in most countries, and requires manifestly necessary licenses and rules, which cannot be shirked and require deliberately obstructive hurdles. Furthermore, the rules are such that no human embryo in those labs can be viable and must not progress beyond fourteen days after fertilization.

But with access to human embryos, and the availability of gene editing, for the first time in history we have unprecedented potential control over the very basic aspects of a human life.

Let's circle back to the very beginning of this book, when my phone was buzzing overnight in November 2018. The Chinese scientist He Jiankui had announced his illegal experiment on the two baby girls he had genetically modified in an attempt to inoculate them against HIV infection.

The background was this: there is a gene present in everyone called *CCR5*, which, as with so many human genes, does many things in our bodies. The protein* encoded by *CCR5* is one of a large family broadly and boringly called a seven-(pass)-trans-membrane G-protein coupled receptor.† There are at least eight hundred different types of this protein in humans, doing myriad jobs to sustain our lives—including turning photons into

---

* Remember that genes encode proteins. The gene is merely the coded message, and the protein is the translation that performs a function. A gene and its protein tend to have the same name, though we specify the gene by writing it in italics while the protein remains in roman type.

† What this means, if you are really interested, is that it's a large molecule that straddles cell membranes, with one end on the outside and the other on the inside, and loops in-out-in-out seven times. In a very general sense, therefore, this family of proteins act as way stations for information to pass in and out of cells.

visual information, detecting smells, maintaining hydration in cells, and a whole lot more.

The protein CCR5 plays a big role in our immune system. It sits on the surface of white blood cells, the sentinels that patrol our bodies hunting for infectious agents that wish us harm in prolonging their own existence. The immune system is pathologically complex, terrifyingly so even for the bravest biologist, and white blood cells come in a variety of flavors, with glorious names such as macrophages, T cells, dendritic cells and eosinophils, each of which do different things in protecting us from pathogenic attack. CCR5 works on the surface of all these cells, and thus is involved in a whole range of molecular interactions throughout our bodies, from limiting the damage of strokes in the brain to inhibiting the growth of tumors. As with so many human genes, its role in inflammation is not well understood.

One thing that has elevated interest in *CCR5* since its discovery in 1996 is its role in HIV infection. The normal version of the CCR5 protein sits on the surface of some immune cells and acts as a gateway for certain molecules. HIV has evolved to take advantage of that opening and uses it to jimmy open the cell membrane and infect cells to enact its own genetic program, and in doing so, cause AIDS. As with all genes, different people have subtly different versions of *CCR5*, but there is one rare variant where there is a chunk of the gene missing: thirty-two letters of DNA for some unknown reason have become deleted, and for that reason the variant is known as Δ32.

Because there's a chunk missing, the protein made by *CCR5* Δ32 is smaller, and as a result, HIV cannot use the gateway, and can't infiltrate the cell. What this means is that people with *CCR5* Δ32 are immune to HIV infection. You have to have two Δ32 versions to have this immunity, and this is rare: about 1 percent of people descended from largely White European populations have two copies, and none have been found in people from East Asian, African or Native American populations. We don't know why this mutation evolved and survived, but some scientists have speculated that it may have emerged in response to the Black Death around seven hundred years ago. Then again, not every variation in our genomes has to have been selected to survive. Much of it is just stuff that randomly changed, and didn't kill us.

The father of the twin Chinese girls is HIV positive. He Jiankui's idea was to use the current techniques available to us to give the man's children immunity to HIV. Using IVF, He took fertilized embryos and, deploying CRISPR, attempted to delete the thirty-two DNA letters in the normal *CCR5* gene to turn it into the Δ32 version. He then implanted two of those genetically modified embryos into their mother, and the babies were delivered nine months later: Lulu and Nana.

This is illegal. Under Chinese law and international agreements, no work on human embryos is permitted if they are older than fourteen days, or can come to term. But set this aside for a moment and park your righteous moral indignation. On hearing this news, my moral horror was matched by a deep and expert scientific skepticism to the very possibility of what he proposed and, from his own announcement, *failed* to enact. Some of the details are not fully known but based on the available evidence, including data that He himself presented, I am fairly confident about the following things. The first is this: the intended gene editing failed. Professor He did not successfully create the $\Delta 32$ version in either embryo. Instead, in the one that became Lulu, a run of fifteen letters of DNA was deleted. In Nana, some DNA was added and other parts deleted. When presenting his data, He suggested that this would break both copies of *CCR5* and thus provide immunity to HIV infection. But that is completely unknowable. The mutations in the embryos are unknown to both nature and science, and we have no way of understanding what effect they will have. Any sane person would have abandoned this procedure right there. Professor He had failed to do what he set out to do. But he went on and implanted the embryos anyway. This is one reason that this whole affair qualifies as human experimentation rather than a therapy. The pregnancies continued, and two girls were born who had been experimented on. The other reason is that the proposed edits were not treating a disease, or even the risk of a disease. It was a genetic prophylactic, a lifelong DNA condom that would in theory release two people from the risk of infection that most people avoid anyway.

The fact that the desired edits didn't happen is not a big deal scientifically. CRISPR is accurate and precise, but it's certainly not perfect. Any scientist worth their salt would do a procedure dozens of times to ensure the edit they were aiming for. Furthermore, there's another phenomenon with CRISPR, well-known but not well understood, called off-target edits. Even though with CRISPR you load it with a sequence that should target the gene you are aiming for, there's a lot of DNA in a cell, and CRISPR can latch on to other bits of the genome and edit away. A decent scientist would attempt to find out if this has happened, because introducing random mutations in a genome is what causes cancer, or breaks genes, or any number of other undesirable stochastic outcomes. He Jiankui says he did this, but the evidence for his claims has never been seriously scrutinized or peer-reviewed.

There's even more garbage in this genetic cesspool. He attempted to perform the edits on the fertilized egg as a single cell, and therefore whatever changes were introduced would in theory be present in every single cell in the children's bodies. But in Nana, the procedure seems to have taken place at a fractionally later stage of pregnancy, meaning that whatever edits took hold, not all of her cells would have them, and she would be what is known as a "mosaic," with different cells bearing different types of genetic lineage. He says that he explained this to the parents, and they elected to go ahead anyway. Under normal conditions, this should not have been their choice to make. He claims he had ethical approval from the Shenzhen HarMoni-Care Women and Children's Hospital. The hospital says no such records exist. The full details of the repercussions and aftermath of the announcement remain vague. We know that He himself was jailed and fined—three million yuan (around $475,000) and a three-year sentence (though he's since been released). What happened to the girls, their health and well-being, is not currently in the public domain.

The moral and legal violations committed by He Jiankui in

this incident are largely unarguable. But the technical problems should not be surprising. There have been a handful of published attempts to modify disease genes in human embryos using CRISPR in the last few years, though all in accordance with the law, in embryos that were destroyed before the internationally agreed time limit of fourteen days post-fertilization. The results have been mixed.

One of the first experiments, in 2015, used embryos that had been fertilized in vitro with two sperm, as a means of ensuring their unviability. A team of Chinese scientists from Sun Yat-sen University in Guangzhou used CRISPR to cut a gene and observe its repair. The process was slightly successful, but they found that only four of the fifty-four attempts had succeeded, and even the four successfully edited embryos were mosaics, harboring some cells that had been corrected and others that had not. Two years later, another Chinese team tried to correct mutations in two genes that cause diseases in viable embryos and got three out of six. At about the same time, another team, this time a U.S.-Chinese collaboration, did the same with a gene called *MYBPC3*, mutations in which can cause hypertrophic cardiomyopathy, the most common form of sudden heart attack in otherwise healthy people. They did manage to correct the mutation, but off-target mutations and mosaicism remained a problem.

That's roughly where we are now. There have been a few other experiments, each inching toward better efficiency with fewer problems. All the genes targeted so far have been ones in which mutations are precise—mostly single-letter changes—and they are largely monogenic, that is, one gene responsible for the majority of the disease. The off-target mutations will always be a problem until the CRISPR system can be shown to be so precise that this never happens. If this technique is ever to be successful as a therapy, mosaicism needs to be eradicated. In a ball of cells, we don't know which ones will become which tissue because those decisions haven't been taken at that early stage of development. If you want to fix a heart defect, then the corrected mutations need to be present in heart tissue, and if not all the cells in an embryo have that correction, then there is no guarantee your edit will be of any use whatsoever. These are the issues we face before considering the next step. These are the issues that were ignored by He Jiankui.

I expect these problems will be overcome in the not too distant future. They are technical and fiddly, which for the most part in the biological sciences simply means that more meticulous and thoughtful work needs to be done. If, or rather when, those barriers are satisfactorily hurdled, we face the ethical question of whether or not to allow an experimental procedure on an embryo with the intention of implanting it, growing to term and being born. I don't know what I think about this. As a procedure for gene therapy, it has its merits, but is costly and laborious, and it has a risk that is greater than of embryo selection following PGD while the outcomes are the same. As a protective measure, a medical prophylactic, I can't really see any justification. It's a version of the type of experimentation that you might see in a superhero comic, only significantly more feeble and much less dramatic.

A SCIENTIFIC CREED (PART TWO)

The saga of He Jiankui is one of many reasons that talk of the return of eugenics is spreading. The technology is available to select embryos that are free of certain diseases. Gene editing technology is beginning to be available to modify genes with pinprick accuracy. Our knowledge of genetics, and particularly polygenic behaviors, is such that we're mapping the pinpricks.

If parents were to select an embryo based on eye color, or the probability of the resulting child's being nudged a fraction of an inch toward a slightly higher IQ, you aren't actually providing any new genetic information, you are merely choosing. You're choosing DNA that was present in the parents, but this way, you're reducing—albeit by a fraction—the role of chance in the genetic makeup of your kid.

This is not a new idea. In 1980, a businessman called Robert Graham set up the Repository for Germinal Choice in Escondido, California, which ran for nineteen years and provided sperm that resulted in the creation of around 220 people. Graham's claim was that all the sperm donors had been carefully chosen from Nobel Prize winners and other senior academics, and he defended his business with a line straight out of the most basic eugenics playbook. "The better the human gene pool, the better the individuals who will come out of it," he said, "and the poorer the human gene pool, the more useless and detrimental individuals will come out of it." The business ended in 1999, two years after Graham died, and the facts seem a little bit less certain than the public relations. The only Nobel Prize laureate confirmed to have donated was the American physicist William Shockley, who won his gong in 1956 for groundbreaking work on semiconductors.

Shockley's donation was no coincidence. Since the 1960s, he had expressed an ardent interest in eugenics as a means of fixing that age-old panic of the Victorian eugenicists: he believed the inevitable decline of American society was a result of less-intelligent people outbreeding the smart ones. He proposed voluntary sterilization for people with IQs less than 100 (by definition, that is 50 percent of any population). He was also adamant that the not so smart ones were bound to their low intelligence by genetics, and that African Americans were unsalvageable for that very reason. "My research," he said, "leads me inescapably to the opinion that the major cause of the American Negro's intellectual and social deficits is hereditary and racially genetic in origin and, thus, not remediable to a major degree by practical improvements in the environment."

It's not clear who was fathered by Shockley's sperm, and most parents withdrew from follow-up studies. But of the known children who were conceived via this business, most seem fairly normal.

Nowadays, PGD is a treatment available in many places but proportionately to only a few. Germline gene editing—that is, altering the genetic code such that the edit is carried in all cells, including sperm and egg—is currently experimental, and it may yet translate into real treatments one day. If successful, those treatments have the potential to eradicate a disease from the germline of that individual and all their descendants.* But germline editing will not affect populations, nor shape society significantly. Furthermore, it will only be available to a tiny minority. That is not to say that the selection of embryos free from certain diseases should not happen, but evaluation of germline gene editing must be done on a case-by-case basis, with the benefits and risks assessed in a thorough way. When these standards are met, as should be determined in consultations involving a wide range of people with expert advice to hand, then and only then might we take the next step.

Dozens of academic papers discussing the morality and ethics of PGD have been published in the last few years, and the numbers are increasing. Some of them include thoughtful discussions of this phenomenon. But there are also plenty of articles taking superficially provocative and iconoclastic positions that seem intended just to enrage and provoke on social media, when actually they're mostly just semantic arguments about definitions of words. Ethical discussions about research into human embryos and genetics occur in every lab as standard, and are a prerequisite for the work to take place. In my experience, these essential and thoughtful processes are largely

---

* Indeed, there is one example where this has already been approved. In mitochondrial replacement therapy, a discrete loop of DNA that bears a disease mutation is replaced by one from a donor in embryos, which if they grow to term, will have the healthy version in all their cells and will not suffer from the terrible diseases that otherwise would have threatened their lives. The procedure has been approved in the United Kingdom, but it is not public whether any children have been born via this method.

unaffected by the intellectual posturing of academics who aren't really involved, but enjoy a scrap on social media.

Nevertheless, as with He Jiankui, there are always going to be people who are keen to take the next step, either with or without the consent of society. Stephen Hsu is one such person, an American former physicist and administrator at Michigan State University* who became fixated on questions of genetics, embryo selection and intelligence. As a regular media pundit, Hsu frequently talks about the predictive power of genomics for many traits, and often about intelligence, and how it will soon be possible to select for significantly smarter children.

Hsu and a Danish collaborator named Laurent Tellier founded a company that sells a forecasting service for prospective parents undergoing IVF. Genomic Prediction Inc. offers tests for eleven conditions and provides a scorecard of risk for a handful of diseases. It's different from standard PGD for specific diseases with single-gene roots, such as cystic fibrosis (CF) or Huntington's, as the company's service is for polygenic disorders—including diabetes, coronary heart disease and four types of cancer. Embryo selection for traits (as opposed to dis-

---

* In June 2020, Hsu resigned from this post at the behest of the MSU president, after a petition and campaign led by students accused him of expressing scientific racism, sexism and scientific dishonesty (in failing to disclose his financial stakes in Genomic Prediction Inc. in published papers) and of being a eugenicist, allegations he vigorously denies. Hsu briefly became a cause célèbre on the grounds of academic freedom, but having resigned his vice-president post, returned to a tenured faculty position, where he remains today.

eases) is not permitted in most countries in the world, with the notable exception of the United States. In most European countries, the application of PGD is regulated by the state, and their governments have placed restrictions on what criteria can be used in selecting embryos for implantation. In America, there are currently no federal or state laws that limit these criteria. Elective (that is, nonmedical) sex selection during IVF occurs in about 9 percent of cases. But without specific legislation that limits what genetic makeup can be chosen by parents, the possibility of selecting for polygenic traits is currently wide open in America.

The current Genomic Prediction Inc. scorecard on sale does not include any traits, only diseases—though one result is for a genomic prediction of extreme intellectual disability. The company does not offer selection for superior intelligence, but that is clearly on Hsu's mind. In a 2014 essay entitled "Super-Intelligent Humans Are Coming," he invoked the age-old eugenics canard of agricultural breeding: "Broiler chickens have increased in size more than four times since 1957. A similar approach could be applied to human intelligence, leading to IQs greater than 1,000."

That's pretty nutty, mostly because the way the IQ test works, it is simply not possible to get a score of 1,000, just like you can't score 1,000 in ten-pin bowling. Hsu also speculated that some countries might not have the same attitudes to these putative services. "If the HFEA* decides that it's not right for the UK, I

* The U.K. Human Fertilisation and Embryology Authority.

will respect that," he told journalists in 2019, but if that happens, "rich people from the UK will fly to Singapore."

Or indeed America. In July 2021, Aurea Smigrodzki celebrated her first birthday in North Carolina. As far as we know, she is the first child to be born where polygenic risk scores were used by her parents to select a specific embryo, via Genomic Prediction's service. Rafal Smigrodzki, her neurologist father, likened this to a duty of care to a new life, telling Bloomberg News in September 2021, "Part of that duty is to make sure to prevent disease—that's why we give vaccinations. And the polygenic testing is no different. It's just another way of preventing disease."

The science on which the service is based is not unproblematic, and almost certainly of very limited practical use—I'll go into some details in the next section—but Hsu's voice can be heard in the public discourse talking openly about embryo selection for intelligence in a way that many people legitimately find disturbing.*

To my mind, this situation is slightly reminiscent of the

---

* Genomic Prediction is not the only company on the market offering such services. Orchid Health offers a similar service, as does MyOme, whose tag line is "Harnessing the True Power of Genetics." In documents available online, MyOme also offer polygenic risk scores on embryos for several behavioral traits as part of a study program, including "education attainment," "household income," "cognitive ability" and "subjective well-being." But in reference to those traits, it also contains the disclaimer "The likelihood of the disease or trait manifesting itself may not be any greater than it happening by chance," which makes me wonder what would be the point in paying for this service in the first place.

genesis of eugenics, and the key question of how an esoteric academic idea became policy in the most heinous crimes of a century. Hsu's influence is certainly not on this scale, but ideas can trickle a lot faster these days than in the nineteenth century. In 2014, Hsu caught the eye of a British civil servant called Dominic Cummings, who saw him talk at a science/tech meeting called Sci Foo, held at Google's base in California.* Cummings would go on to become the key architect of Brexit, Britain's withdrawal from the European Union. He then became the manager of the Conservative Party's election victory in 2019, and finally the chief adviser to Prime Minister Boris Johnson. Back in 2014 though he was relatively unknown, an odd policy adviser with a voluminous blog.

On that blog in 2014, Cummings writes at length—and largely accurately—about genetics and intelligence, and the possibilities of embryo selection. He cites the work of Hsu with reverential tones. He inadvertently predicts what Hsu would go on to do: "Once the knowledge exists, it is hard to see what will stop some people making use of it and offering services to—at least—the superrich." Cummings also raises the all-important ethical question:

---

* Sci Foo is a three-way collaboration between Google, where it was held, the journal *Nature* and the tech publishing company O'Reilly. It was intended to be a private meeting of fine minds with great ideas in a range of fields, with each company inviting one-third of the participants. For the record, I played a part in organizing this meeting during the time I worked as an editor at *Nature*, though I had already left the company by 2013, and played no role in this particular meeting.

Once we identify a substantial number of IQ genes, there
is no obvious reason why rich people will not select the
egg that has the highest prediction for IQ. This clearly
raises many big questions. If the poor cannot do the same,
then the rich could quickly embed advantages and society
could become not only more unequal but also based on
biological classes.

There is an obvious reason though. IQ correlates with many
things, some considered positive such as income and longevity, and other not so desirable, such as mental health problems.
Our current knowledge of genetics is not sufficient to disentangle why.

For all my criticisms, Cummings positively cites a BBC radio
series that I wrote and presented on this very subject, so it
appears that I may have inadvertently influenced his thinking
on this subject.* So it goes.

Some of us read that blog post at the time and thought nothing much of it. Some wondered why these ideas were so central
to an influential government adviser, when we already know
how to improve the intelligence of populations with better education, health care and access to physical exercise, without having to fantasize about tinkering with genes that would only be
accessible to a minuscule minority.

Hsu, on his own blog in 2019, reciprocated the fraternal ado-

* It was a three-part series made in 2014 called *Intelligence: Born Smart, Born
Equal, Born Different*, which is still available online.

ration that Cummings had previously shown him, and posted a picture of the two of them at the door of 10 Downing Street. Whatever the moral rights, wrongs, scientific accuracies or fudges of this saga, these ideas were being entertained in the heart of the British government, by the man who, for a time, whispered in the ear of the prime minister.

Moreover, Cummings's blog created a space and an inspiration for dark musings about the potential for genetic selection. A young data analyst named Andrew Sabisky, who was later recruited by Cummings to work for the British government on unspecified projects, commented on the 2014 Sci Foo blog:

> This naturally leads on to the point that one very good way to retain human control over technology—and to think up better ways to ameliorate its negative consequences— is global embryo selection. . . . One way to get around the problems of unplanned pregnancies creating a permanent underclass would be to legally enforce universal uptake of long-term contraception at the onset of puberty. Vaccination laws give it a precedent, I would argue.

Legally enforced sterilization was, of course, central to the legislation wherever there have been eugenics policies. I find Sabisky's views not only deeply troubling but also wrong— global embryo selection couldn't possibly work unless you're in a mad science fiction dystopia—was he really suggesting that all babies should be produced via IVF and embryo selection? I think most people prefer the old-fashioned way.

Everyone is entitled to their opinions. What I object to is that they might be aired by unaccountable unelected mandarins in the heart of a democracy. Sabisky stepped down promptly after public exposure of his musings, and also the discovery of his attendance at an annual meeting in 2017 of controversial science-cosplaying cranks and racists about intelligence research. That meeting, the London Conference on Intelligence, was for a short time an annual get-together for a small cabal of researchers, writers and oddballs for whom race and eugenics are the enduring passions of their lives. The intelligence researcher Richard Lynn was there, a man with strong White supremacist links and a history of producing highly questionable racist work. Also in attendance was the current head of the Pioneer Fund, an American foundation set up by Harry Laughlin and other keen eugenicists in 1937 to study and promote ideas about scientific racism and eugenics. Laughlin, if you recall, was Charles Davenport's deputy, and wrote the legislation in the United States that was used as the template for the 1933 Nazi sterilization legislation. The Pioneer Fund, among its many activities, maintained close ties with the Third Reich, including acts such as funding the distribution of films to garner public support for Aktion T4. The fund still exists today, mostly gasping for air in the fetid corners of the internet, hobbling along as an ancient artifact of its profoundly racist past, and publishing journals that bear superficial resemblance to credible academic or scientific work but are really of interest only to cranks, self-styled heretics and racist fools.

People such as Dominic Cummings, Stephen Hsu, and other shallow-thinking provocateurs are perfectly entitled to their opinions, but it should be emphasized that opinions is all they are. Of course, there is a role for iconoclastic thinking and incendiary thought experiments as society wrestles with new ideas, or even old ideas dressed up in new science. The press loves a heretic, especially one who can set fire to things and then walk away to look for the next pile of kindling. It is more problematic when the arsonist is in the heart of government, but is ill informed and, worse, spellbound by a science that is barely understood by experts, let alone tourists. The label gets flung around as an insult, but I don't think these people are eugenicists, at least not in a comparable sense with the enthusiasts of the prewar era. However, if your understanding of a science extends only to the point where your political preconceptions are supported, then you are an ideologue, not a scientist.

Because some people pontificate about enhancing certain characteristics in people, or reducing undesirable traits, we face similar debates to those that occurred a century ago, when eugenics was enjoying its popular—though not unopposed— heyday. All those tools at our disposal and all that knowledge that we now have at our fingertips certainly gift us the potential to change the genetic makeup of future generations, and people and maybe society. But change does not necessarily elicit control. To understand why, we need to go to graduate school.

TWENTY-FIRST-CENTURY GENETICS 101

We have identified hundreds of points in our genome that contribute to the genetic basis of intelligence. We can tweak DNA and select embryos. All these things could theoretically be combined. "Should we do what we can?" is the ethical question that arises in such conversations, but set that aside so we can deal not with the morality of genetic modification but the scientific reality: is what is being discussed even possible? To do that we need to get back into the nuts and bolts of genetics and sex, and update the basics of heredity.

Earlier, we discussed how the idea of individual traits can be passed from parent to child, and how some genes—notably the first disease genes discovered—are largely caused by a single fault, but other traits, such as intelligence or schizophrenia, enjoy the contribution of dozens of genetic variants of very small effect, in aggregate and in concert with the environment. Traits are encoded in genes, which are made of DNA. Genes make proteins, and all of life is either made of or by proteins. A gene can be defined as a specific piece of DNA that provides a coded instruction to make a protein. We have around twenty thousand such genes in our genome, and each of these encodes a protein that does something in our bodies. Keratin is a protein that provides much of the structure of hair. Hemoglobin is a protein which forms a quartet that holds an atom of iron in its middle, and this transports oxygen in our red blood cells. The melanin that pigments our skin is not a protein itself but is produced in a biochemical pathway that is controlled by proteins. The genes that encode all of these proteins are inherited from mothers and fathers, and from nowhere else, which explains why people tend to resemble their family more than random strangers.

But of course, we aren't clones of our parents, nor are we solely blends of their traits. We are shuffled and spliced as sperm and egg are formed, and in all that remixing there is novelty, just as Philip Larkin* wrote:

They fill you with the faults they had
And add some extra, just for you.

What version of whatever gene you get from your parents is important not just because of what its protein does. How they are inherited is also relevant. You have two copies of each gene, one from each parent, but which of these is selected for action is also not straightforward. Sometimes, one gene can be dominant over the other, which we call recessive. One of the genes heavily involved in eye color is called *OCA2*; one version causes lower pigmentation levels in the iris and is therefore part of the genetic ecosystem that codes for blue eyes; another version codes for higher melanin pigmentation and is associated with brown. Blue is recessive, and brown is dominant over blue, meaning that if you have inherited a brown version from one parent, and a blue from the other, you will probably have brown eyes. If you have inherited brown versions from both parents, you will have brown eyes. Only if you inherit blue versions from each parent will you have blue eyes. You learned that in high school.

In disease genetics, these rules are also of paramount

---

* "This Be the Verse," from *Collected Poems* (Farrar, Straus and Giroux, 2001).

importance. Take the three diseases I've already discussed: cystic fibrosis, Duchenne muscular dystrophy and Huntington's. Cystic fibrosis is a recessive disease, meaning that if you have inherited one broken copy of the gene *CFTR*, you are a carrier of the disease, but have no symptoms. To have cystic fibrosis, you must have inherited two copies, one from each parent. Another example of a recessive disease is sickle cell anemia. If you have a single faulty beta-globin gene you may get sickle cell trait, which is a serious condition but manageable, but if you have two copies you will get sickle cell anemia, which is much worse. These two examples, like blue eyes, are recessive. Huntington's disease is caused by a dominant mutation, meaning that you need only one copy of the expanded *HTT* gene to get the disease. Duchenne muscular dystrophy is what is known as X-linked: the gene *dystrophin* is on the X chromosome, and if a girl has a faulty copy, her second X may well compensate. Boys have no such insurance, as they have only one X, and this is why almost all sufferers of DMD are boys.

When we teach this, we tabulate the potential results for a child in what is known as a Punnett square, one parent's genes on one axis, the other on the other. The combinations of potential children fill the table out. As mentioned, this very handy tool was invented by Reginald Punnett, the Edwardian geneticist who so confidently attributed feeblemindedness to a single gene at the International Eugenics Congress in 1912, and who argued that society therefore should act to breed it out of existence. He was wrong on both counts.

These basic rules are the ones laid out by Gregor Mendel and his pea-plant experiments in nineteenth-century Moravia. A cross between a purple-flowering pea plant and a white one did not produce a pink one, but ratios of either purple and white flowers in their offspring. The gene for purple flowers is dominant over white. But this is only the edge of a very complex picture. There is also co-dominance, where versions of genes from both parents are expressed in the offspring. There is epistasis, the process by which two different genes at different places in the genome can have a combined effect on a trait. Sometimes this can be additive, in that two mutated genes can contribute to the same problem and exacerbate it, or they can mask it. There are dozens of complexities within the basic Mendelian rules of inheritance. In reality, they're more like guidelines.

It took the Human Genome Project to reveal this. One of the things that we had fumbled over was that we had hugely overestimated just how many genes humans have. A common assumption had been that there were specific genes *for* everything—a gene for eye color, a gene for every disease or trait. What the HGP revealed is that 20,000 protein-coding genes is the full complement of a human genome, and it is not nearly enough to satisfy that model. It's about the same number as a cat has, but way less than a banana (around 36,000). The Lego set of Rome's Colosseum has 9,036 pieces.

So, we began to work out a model in which genes work in complex networks, and some genes do multiple things at different times and in different tissues from conception onward. Years ago, I worked on genes in the eye, one of which also had functions we don't really understand in the liver and in the brain. Another did different things in the eye at different stages of embryonic development, and this depended on other genes being active at the same time, all interacting in a delightfully choreographed but intricate dance. When it all goes according to plan, what started out as two small indentations in one end of an embryo blossom into the most exquisitely designed organs in the body, elaborately evolved to turn photons into perception. When it goes wrong, children are born congenitally blind, another condition specifically targeted by eugenics policies.

What we know about human genetics is immense, and in just twenty years that knowledge has expanded exponentially. What we don't know is much more. I tell you these things not to shame my colleagues for not yet having a clearer picture of our nuts and bolts, but to praise them for revealing that we, the most complex organisms in existence, could not be explained only by simple laws derived from pea plants. Instead, we bathe in a world of unending data, picking at the frayed edges, eking out the details of a system that has been evolving for four billion years longer than we've been looking at it. Biology is messy. When it goes right, you get a healthy baby. When it goes right again, you get a completely different baby. It can go wrong in an immense number of ways, ranging from the trivial—a crooked tooth, a birthmark—to the troublesome, all the way to the life-defining, life-threatening or lethal.

The trouble with the way we teach—and tend to think about—genetics is that it is rooted in the simple Mendelian sense of one gene for one trait. The trouble with using blue eyes or red hair as examples to teach human genetics is that they're not very true.

Eye color is a spectrum from very pale blue to almost black. Let's go back to that example of the gene *OCA2*, which is involved in the pigmentation of blue or brown eyes—we know that the brown version is dominant and the blue one is recessive. When we look really closely, this model—which was first described by Charles Davenport in 1907—accounts for only about three-quarters of the variation in blue and brown eyes. Remember that you need two blue versions to have blue eyes? Well, only 62 percent of people with two blue versions have blue eyes, and 7.5 percent of people with two brown-eye versions of *OCA2* have blue eyes as well. Several other genes (with soul-draining tax-code names such as *TYRP1*, *ASIP* and *ALC42A5*) also have significant roles in pigmentation distribution in the iris and profoundly affect the eye color of individuals. You might think that this particular trait is old news, because it's so obvious to look at and because we teach it in the very basics of human biology. But it's still an active area of research, and in March 2021 a colossal study recruited 193,000 people from ten populations to try to get to the bottom of eye color, using a GWAS approach, and found 124 independent associations arising from 61 discrete genomic regions, including 50 previously unidentified. We find evidence for genes involved in melanin pigmentation, but we also find associations with genes involved in iris morphology and structure.

Now, you know this intuitively, because blue-eyed parents can have brown- or even green-eyed children. Even if you come from a family where all eyes are the same color, they're really not when you look closely. This trait is not binary. Eyes can have multiple pigments within them, either in the striking forms of heterochromia, where discrete patches of different colors are visible, or even different colors in each eye. The genetics of eye pigmentation is beautifully complex, and not fully understood.

And yet we teach it as gospel, a model of monogenic determinism that was the mainstay of Davenport's principles of eugenics. Even the modern popular commercial genetics tests that millions have paid for are clear on how unclear the genetics of eye color actually is, compared with how we teach it. Though I have spent a career expressing deep skepticism about those new commercial genetic tests that purport to reveal your ancestry or physical traits, there is useful information within those results that reveals the muddle of human genetics. 23andMe examines the *OCA2* gene as part of its service, and my report says that I've got a version thrillingly labeled rs12913832. I have dark brown eyes, something I know not by having paid to have my spit analyzed but because I own mirrors. According to 23andMe's client database, 31 percent of people with this version of *OCA2* have dark brown eyes. The other 69 percent do not. If you were to dig me up in ten thousand years' time, and managed to get DNA out of my old bones, and you wanted to know what color eyes I had, then you'd have a one in three chance of getting it roughly right at best. Be aware of those odds when you next read of the physical description of a long-dead person who has been reconstructed based on their ancient DNA.

Or let's look at ginger hair. The gene that influences hair pigmentation when it comes to redheads is called *MC1R*. At least seventeen versions of *MC1R* have been identified that are closely associated with red hair, each subtly changing the shape of the protein it encodes. Because red-hairedness is a recessive condition, you must have two copies of a redhead version to have red hair—the range of mutations goes some way to explain why there is variation in the many shades of ginger. $MC1R$ prompts the production of melanin, but there are two types: eumelanin (which is browny dark) and phaeomelanin (which is redder), and redheads have more of the latter than the former. Red hair is a standard recessive Mendelian trait.

Except that it's not. A study similar to the one mentioned above about eye color zoomed into the genomes of 343,000 people in 2018 and found that much of the variation in red hair color was explained by the known versions of $MC1R$, but as was discovered in this study, "most individuals with two $MC1R$ variants have blonde or light brown hair" and blonde hair "is associated with over 200 genetic variants." Most people who have two ginger variants in their $MC1R$ gene do not have ginger hair. I carry one version, yet have black hair, except for the occasional bright ginger bristle in my beard. We don't currently have an explanation for red hairs in otherwise nonginger men's beards. The genetics of hair pigmentation is beautifully complex, and not fully understood.

Yet we teach it as gospel, a model of monogenic determinism that was the mainstay of Davenport's principles of eugenics. And this ginger-hair genetics study included only White British people. Imagine how much more complex it will get when we look at the majority of people in the world today.

Eye and hair color are highly heritable traits. The impact

of the environment on determining each is extremely low, almost nonexistent. They are traits controlled by DNA, inherited from your parents. They are the traits that a eugenicist might delight in, because we know how insignificant nurture is compared with nature. However, only now in the third decade of the twenty-first century are we beginning to understand how they work.

This is why I never get too exercised in conversations about "designer babies," or popular Hollywood films such as *Gattaca*.* When people start anxiously or glibly opining about gene editing for designer babies or selecting embryos for blue-eyed children, they're not really talking about our contemporary understanding of genetics. Instead, they are relying on a hugely outdated or a never true version of altering heredity that is pretty much impossible. If you were to select an embryo for eye or hair color based on the versions of the genes most closely associated with blue eyes or red hair, you'd have an expensive gamble, and a hard-to-calculate chance of getting what you want. Our current understanding of genetics does not allow for control of even the supposedly simplest traits.

When it comes to genetic diseases, which have very serious

---

* The 1997 film starring Ethan Hawke and Uma Thurman. It's set in a near future in which a very Galtonian form of genetic selection plays the significant role in determining which jobs and status in society people hold. It's okay, I don't love it, but it had a rather good advertising campaign, with fake newspaper ads offering "Children made to order," and the tag line "How far will you go?" Images from *Gattaca* now feature in the slide decks of 87.62 percent of talks about human genetics and eugenics.

implications, then the game is not so funny. Even with the best understood genetic diseases such as cystic fibrosis, the picture is very messy. Around two-thirds of people of White European descent with CF have the same mutated version of the gene *CFTR*. The ΔF508 version makes a protein that is truncated, like a sentence that is several words too . . . The truncated CFTR doesn't work well in its role for moving salts in and out of the cells that line the lungs, and the result is a thick mucus that inhibits breathing. This genetic etiology may sound confidently clear, yet more than fifteen hundred other mutations in this one gene have been described that can also cause CF. Furthermore, even for the majority of people who have the common ΔF508 version, the severity of the symptoms is far from identical, and in 2015, five genetic modifiers were identified that were not part of the *CFTR* gene at all but had significant roles in how severe the disease was. How these genetic variations affect the patients is not known.

These are the traits and diseases that we know the best, the ones we teach as the basics in schools. And yet they are riddled with complexities that to my mind preclude gene editing interventions as a means of eugenically removing or altering them from an individual and from future generations. Because of the nature of our DNA, the permutations are effectively infinite. Yes, variants in *MC1R* are part of the cauldron that cause red hair, but the variants are endless. Yes, versions of *OCA2* significantly influence eye color, but the variants are endless. Yes, the vast majority of CF patients have the same mutation in a single gene, but thousands do not, and to understand the endless

variants in this disease and thousands of others is the ongoing
work of some of the best scientists in the world.

For complex traits, obviously, the picture is only going to
get much, much more complex. Those genetic variants associated with traits or disorders such as intelligence or schizophrenia number in the hundreds, each individual version tilting
the scales ever so slightly toward one end or the other. However many genes are involved—several hundred or a thousand,
according to the most recent studies—each one of them contrib-
utes a tiny fraction of the total heritability that we can measure.

When it comes to intelligence, yes, there are variants in hundreds of genes that load the odds toward slightly greater success
in the metrics of cognitive abilities when we look at populations with similar ancestry. This is good knowledge, and worth
knowing. It takes real expertise to figure out just what is going
on in our genomes, but real chutzpah to suggest action based
on this incomplete knowledge. The predictive value of knowing
the genes involved in height or intelligence (or any trait, for that
matter) is worked out by correlating them with the known outcome in a population—you look at the DNA in adults and match
them to their heights or cognitive abilities, or whatever you're
interested in. We can't do that in the same way for embryos
because this technique has been available for only a couple of
years, and those embryos could still only possibly be babies at
this time. The variants are known because we can look at the
genes in huge populations, hundreds of thousands, from which
statistical significance emerges. But embryos generated during
IVF come in batches of very small numbers, maybe eight or ten,

if you're lucky. A team of American and Israeli scientists modeled the possible utility of embryo selection in 2019, as proposed by people like Stephen Hsu. The results are not very impressive. The researchers used a battery of statistical techniques to work out what range of height or intelligence might emerge during a hypothetical round of IVF. The answers came out at just under an inch, and 2.5 IQ points. The typical error margins around IQ are about plus or minus 5 points, meaning that when you get an IQ score of say 120, it was actually somewhere between 115 and 125 when you took the test. And as for height, well, I can get up to six feet tall with the right shoes and hair product.

So, from a practical point of view, embryonic selection for complex polygenic traits is barely viable. It's a costly, technical procedure for arguably marginal gains that could be applied to a handful of people already undergoing IVF. And let's not forget that IVF is not exactly an easy procedure. Any woman who has been through it will confirm this readily. It's a lot less fun than sex, and people who go through this arduous, emotionally fraught and often physically painful process mostly do it as a last resort to have kids when the standard options have not borne fruit.* It's not fun and it's not a game. You may have noted that most of the people who knock around the idea of embryo selection tend not to be the ones who will have to endure the daily rounds of injections to induce ovarian hyperstimulation, or the needle through the vaginal wall to get access to the ovarian follicle. The people who seem most excited by the idea of

---

* For many lesbians, it may be their only option.

eliciting molecular control over reproduction don't tend to have
wombs at all.

Would I want to select embryos with those variants, or even
edit the genomes of embryos to harbor those variants? No. Not
while the roles of those bits of DNA are poorly understood. Not
when we don't know if selecting for something means you are
inadvertently selecting against something else. One study found
that IQ positively correlates with anorexia, anxiety disorders,
attention deficit hyperactivity disorder and asthma. Though, as
I've argued, the gains might be slight, a nudge taller or smarter,
you may also be nudging that child toward an eating disorder or
other unforeseen health problems.

I would also not want embryo selection when the gains of
those variants are so marginal that they can be overwhelmed by
solutions that are known, and understood, and can be deployed
to populations instead of individuals—things as radical as education for all without privilege, tailored to individual needs.
Things like better nutrition, health care, exercise, welfare. If we
want the betterment of our people—and who doesn't?—we don't
need to turn to a scientific creed that is at best poorly understood. When people preach about the possibility of improving
a person or a nation's cognitive abilities, they almost never have
done the legwork to understand the issue. We don't have the
knowledge or the tools, and the expense would be colossal for
the probability of a gain that could easily be swamped by the
inherently uncontrollable nature of biology and life. Biology is
unruly, and therefore cannot be ruled. I struggle to imagine a
time when this will ever be different.

My caution is not derived from fear, nor from a political stance about liberty or control. It is from a position of expertise. I know that what I've written in the last few pages is technical and possibly difficult for a lay reader. You didn't pick up this book to get an undergraduate university course in human genetics, and I've barely skimmed the surface. But that is kind of the point. The glibness with which nonexperts start waffling on about a new eugenics enabled by current technology is frustrating to hear, because it is only in the technical nuts and bolts of basic biology that we can comprehend how wrong they are, or how improbable or impossible the proposals might be. Or that they are flirting with an idea that in history caused so much harm, without the knowledge to back up the claim that it might be different this time. Just as the first time round, and probably the next time, they simply turn to a science they don't fully understand.

DID EUGENICS WORK?

As I hope you can see, I am extremely skeptical about the possibilities of enhancing or removing genetic traits or diseases from populations using the technologies available to us (or rather, to some) today. Whether these qualify as eugenic interventions is up for debate, but in some ways I don't really care. I think it's reasonable to assume that if these technologies were available to the Victorians or Edwardians or Nazis, then they would be excited at the prospect of changing biology as a means of population control.

The eugenics enacted in the first half of the twentieth century defined cultures and conflict, and became a central pillar of the global stage. Science as policy is never clean or tightly controlled as it would be in a lab, but the mass sterilizations and murders did happen, justified by their scientific beliefs, naïve and political though they were, which means that we can ask questions about how successful they became.

We can look at the actions and extract some kind of information about heredity in populations. Obviously, just as eugenics in the Third Reich merged with deranged, racist, homophobic and hate-filled ideologies, many of those actions will not survive much scientific scrutiny.

Homosexuality cannot be eradicated because it is natural and not maladaptive. Of course, homosexuality is heritable (as per the first rule of behavior genetics), but sexual behavior is not a binary; some people are exclusively gay, others bisexual, some exclusively heterosexual. Some people change their sexual preferences during their lives. Others hide stark differences between their thoughts, words and deeds. There is a genetic component to sexual behavior and that includes homosexuality, but those genes don't make you gay. You don't have to be a world-class geneticist to notice that every homosexual person who has ever existed was the result of heterosexual sex.* If your deranged,

* Or at least the pairing of heterosexual gametes, in the case of surrogacies or IVF.

hate-filled desire was to murder every single gay person in a country, well, first you wouldn't be able to identify them, and second, in the very next generation after this horrific genocide, you'd still have a proportion of gay people, and a rainbow spectrum of all the sexual proclivities that humans enjoy. Indeed, one study from August 2021 cautiously suggests a possible reason for the persistence of same-sex behavior despite its superficially appearing to be something that nature would select against—as homosexual people tend to have fewer children. What was found is that the genetic elements that associate with same-sex behavior also associate with an increased number of sexual partners in heterosexual people. It's a preliminary study, and the data limited to people in the United States and United Kingdom, but if correct, whatever those individual bits of DNA that correlate with an increased probability of homosexual behavior are, they are being preserved in heterosexual people anyway. Killing gay people would not alter the existence of the genetic architecture that comes with gay people.

Very obviously, the largest part of the Holocaust was an intention to purify the Nordic race. This was folly squared, because there is no Nordic race, and there is no such thing as racial purity. The tangled web of ancestry that is the genealogical truth of human history precludes even the vaguest concepts of racial purity, and every Nazi had Jewish ancestors, just as every White supremacist today has North African, southern African and Middle Eastern ancestors, if only they bothered to look at and understand the basics of population genetics. The only thing pure about the Nazis' focus on Jews, and on Roma and Slavs, is its racism. There is no possible scientific justification for this hatred.

But what about the diseases that the Nazis sought to eradicate? Under the 1933 sterilization law, many diseases and conditions were included, and some specified: schizophrenia, bipolar disease, hereditary epilepsy, Huntington's disease, severe alcoholism and others.

As institutionalization grew, the number of patients in psychiatric care had increased radically in Germany in the first few decades of the twentieth century (as it had done in Great Britain and America as well), but with the introduction of the principle of "lives unworthy of life" in 1920, these patients' continued existence was increasingly precarious. A psychiatrist named Berthold Kihn calculated that the number of psychiatric patients in German hospitals was costing the state 150 million Reichsmarks per year, and Hitler agreed that this justified their killing: "It is right," he said in 1933, "that the worthless lives of such creatures should be ended, and that this would result in certain savings in terms of hospitals, doctors and nursing staff."

Around a quarter million patients were killed or sterilized—astonishingly, this represents somewhere between three-quarters and *all* diagnosed schizophrenics in Germany in the years 1933 to 1945. Again, the numbers are not straightforward to verify, but in various studies, the prevalence of schizophrenic patients—that is, the number of people in Germany already living with the condition—was low in the years immediately after the war, lower than other European countries at the same time.

But the incidence—that is, the number of new cases—was high and increasing. A study encompassing the years 1974 to 1980 revealed an average of 59 schizophrenic patients in the Mannheim area of Germany per 100,000. This compared with an average of 24 per 100,000 for eleven studies done in the Netherlands, Italy, Denmark, Norway, Iceland, the United Kingdom, the United States and Australia.

There are many factors that could account for this. Better and more specific diagnoses is one, though this should mean the prevalence would also increase postwar, but it did not, and the diagnostic criteria are the same in other countries. It is not entirely clear what the precise diagnoses were for schizophrenia before 1945, but it is likely that they were actually broader than when criteria were formalized and defined in the modern era. Another possible reason is a changing demographic; in the Mannheim study in the 1970s, 13 percent of the population were immigrant workers. But on inspection it was found that the immigrants had a significantly lower rate of schizophrenia than in the established German population.

Whatever the reasons, the fact that the incidence of schizophrenia had significantly increased within three decades since the last patient was murdered or sterilized shows that the policy was ineffective and scientifically specious. They should have known that: it was already well established that the majority of schizophrenia patients do not have children themselves, and do not have a family history of the disease. Though it is unequivocal that there is a significant genetic contribution to the risk of schizophrenia, the environmental influences are quite capable of amplifying the incidences without regard to its biological heredity.

Ironically, the Nazis' combination of scientific ignorance, naïveté and a steadfast commitment to action may well be the *cause* of the increase in schizophrenia in Germany. Many psychiatric conditions are associated with poverty and living in hardship. It is possible, though difficult to account for, that the state of war, and a broken country in the aftermath of devastating destruction of its people and infrastructure, increased the environmental risks for individuals to develop mental health problems. In at least one scenario, this phenomenon is verifiable.

In September 1944, the Nazis were aware that they were in the endgame and were going to lose the war. In a bitter act of defiant retaliation, they blockaded a large part of western Holland. They embargoed food transports and removed cattle and other food stocks. The canals had frozen by November, and very quickly famine struck. The Hunger Winter lasted until April 1945, when the U.S. Army Air Forces and the Royal Air Force (and other Allies) enacted Operations Chowhound and Manna—eleven thousand tons of food drops that the Nazis agreed not to shoot at. Hundreds of thousands of Dutch were starving, and some eighteen thousand people died in those six or seven months. With rationing records from this time, and good demographic data about who was present and who was conceived during this cruel famine, scientists have followed up with the survivors and their children, and have garnered a huge amount of data on the effects of starvation on people—effectively, an experiment that should never have happened. It is well-known that children conceived during the famine have hugely increased incidences of a whole range of conditions, which include both the physical and psychiatric. Of these, the risk for schizophrenia doubled. The Third Reich's eugenics program was an attempt permanently to eradicate the cause of a disease, but what they actually did was only temporarily quash the symptoms and simultaneously create an environment in which it would flourish. Recall that calculation that psychiatric care was costing the state 150 million Reichsmarks per year in 1933. It's not a linear conversion calculation, but in today's money that is roughly $925 million. The most recent assessment of the cost of treating schizophrenia in contemporary Germany puts the bill at around $13 billion.

Unlike schizophrenia, Huntington's disease does have a single-gene cause, and it is invariably lethal. It's a brutal disease, with rapid degeneration of physical control, starting with fidgety movements, slurred speech and difficulties in swallowing. It also comes with depression and other mental health problems, including obsessive-compulsive disorder and bipolar disease, and after a few years from the first signs, people die without any control or basic bodily autonomy. One of the key characteristics for the majority of people with Huntington's is that symptoms tend to show up only when carriers of a single copy of the mutated gene are in their forties, which typically is after they have had children. A child of a person carrying the Huntington gene has a 50–50 chance of inheriting it, and if they do, they will suffer the disease in adulthood (if they don't die of something else first). This means that until very recently, people often passed on the disease gene to their children before they knew that they had it themselves. Today, we can identify children that have inherited the disease gene from a parent at birth, and in the cases of IVF, in utero, but that was not possible in the days before prenatal genetic diagnoses.

A policy that sterilizes adults who have Huntington's symptoms is already too late to remove the defective gene from the germline and the population. This is one of the reasons that Huntington's disease persists. Part of the first crucial work on the inheritance pattern of Huntington's was done in 1916, again by America's chief eugenicist, Charles Davenport, who claimed that most cases in the Americas could be traced back to three immigrants, probably brothers, from the seventeenth century. If this were true, it serves the eugenics principle that Huntington's disease would not exist at its current frequency if these brothers had not become fathers. It also served the narrative that went hand in hand with eugenics, which is that immigrants are undesirable.

With that relatively simple inheritance pattern, Huntington's path through families led geneticists of the 1980s to the mutated gene that causes it. There are two ways that Huntington's can be eradicated for future generations. The first is that all carriers of the disease are identified before they have children of their own, and using a combination of genetic counseling, prenatal diagnosis, and if necessary pre-implantation genetic diagnoses, only embryos that do not carry the disease gene are allowed to survive. This basically is what happens today in the wealthy West, but owing to the fact that Huntington's is often not identified in people until after they have had children, and this type of intervention is expensive, that ship has already sailed. The second option is to do the same, but sterilize the children who carry the disease gene before they have children. Either way, we erase Woody Guthrie from existence.

Neither option was available to the Nazis, or indeed in America and Great Britain, when sterilization was proposed specifically for Huntington's disease, though never enacted as policy. The policies either enacted or proposed in the time of eugenics would have had no impact on the existence of Huntington's disease in those societies. Why? Because they simply did not understand it. But despite this apparent lack of knowledge (or perhaps disregard for it), Nazi Germany did not stop legislation from neutralizing the genitals of people already dying from a horrible degenerative disease.

We could go through the list of people who under the Nazis, or other countries' eugenic legislation, would have qualified, and similarly assess the efficacy of their plans. And with the benefit of hindsight and a twenty-first-century understanding of genetics, most would come up as unsuccessful.

The conditions that the Nazis focused on could not have been eradicated by the program of euthanasia and eugenics they employed. Just like in the United States, they identified undesirable people and attempted to prevent them from existing in the future. But due to a limited, naïve and wrongheaded understanding of the forces they were trying to control, in killing or sterilizing people, they were not addressing the causes of undesirability, only the symptoms.

DOWN SYNDROME

Down syndrome is genetic but not hereditary and presents a unique case for the question of reproductive control. It's a condition with a single genetic origin, which is not the same as having a single gene as a cause. People with Down syndrome have an extra chromosome 21, which makes it past the genetic checks and balances that normally maintain the number of each (nonsex) chromosome to two. Chromosome 21 harbors about 231 genes, so there is a whole slew of extra genetic information that is active in the developing embryo. It is not well understood why these extra genes cause Down's, nor why the extra chromosome makes it through cell division. The probability of having a baby with Down's increases with both maternal and paternal age, but more babies with Down's are born to younger women—which reflects the fact that younger women have more babies generally. There is a huge variety of symptoms and severity of those symptoms, but broadly, people with Down's have a range of differences from people with the typical number of chromosomes, which at birth are mostly physical—wide eyes and a flatter face, small hands and feet, and a single crease across the palm. There are also a range of medical issues that come with that extra chromosome and these include heart problems, early-onset Alzheimer's and a life expectancy that is at best in the sixties. Most people with Down's have lower than average IQs.

So, while Down syndrome is genetic, studies on the fertility of people with Down's are few and far between; there are a handful of cases where men with Down's have fathered children without it. This whole area is mired in ignorance and stigma, including a persistent negative cultural attitude about people with learning disabilities having sex at all, though people with Down's certainly do experience sexual desire.

Ableism was an explicit driver of Nazi eugenics. Under the umbrella of Aktion T4, registration of infants with perceived abnormalities—including what was then called "Mongoloidism"—was decreed by the state, and those babies and toddlers were removed under the pretense of being rehoused in state-run facilities. These centers did not in fact exist. The children were marked for *Gnadentot*—"mercy killing"—and parents were sometimes told that their children had died of pneumonia, the bodies cremated to prevent further infection.

A new question arises today, and this generates a fresh ethical dilemma that is incredibly hard to resolve. Women in many countries are choosing to have babies later, a pattern that correlates with education levels in society but which increases the chance of Down's. However, prenatal diagnosis is becoming more and more available to pregnant women, and Down's is one of the first standard checks for the overall health of the fetus and mother. The global incidence of people born with Down syndrome is currently about 1 in 800. However, this number is falling, because prenatal diagnosis for Down syndrome is early, accurate and, unlike the diagnoses that come with IVF, available to most women. In Iceland and Denmark, the incidence of Down's has fallen to effectively zero in the last couple of years, with almost all women opting for an abortion upon discovery that they were bearing a fetus with Down's. In America, about two-thirds of women opt for an abortion, though in

2021, Arizona and South Dakota joined several other states in outlawing abortion when Down's has been diagnosed. Planned Parenthood—the organization set up by Margaret Sanger—is currently seeking to block this legislation.

Many people with Down's have jobs and have degrees of self-sufficiency, though representation and visibility in mainstream culture is scant. Some readers might remember Corky from the early 1990s sitcom *Life Goes On*, the first to have a major character with Down's. More recently, in the era of ubiquitous reality TV, *Born This Way* ran for four seasons from 2015 to 2019, and tracked the lives of seven adults with Down syndrome, with all the joys and drama of their lives. And in 2020, a Netflix science fiction series called *Away* featured a teenage girl with Down syndrome, who is portrayed as capable and sensitive, and develops a close friendship with a nondisabled teen. These depictions in the media may demonstrate a change in the visibility of people with disabilities in public life, and a shift in public perceptions of Down syndrome.

This is, in my opinion, a fantastic, positive step forward for a more inclusive, better and fairer society. However, Down syndrome is a condition that is highly variable, and some people require constant care, and suffer from serious intellectual and health disabilities, and severely shortened life spans. Parents of children with Down syndrome are the ones who have primary responsibility for their well-being, and many say that their children enhance the lives of people around them, with joy and happiness. For others, their experience is significantly harder, especially for those with more severe symptoms, and born into families without privilege and wealth to provide much needed support. Technology has given us the ability to identify a group of people that exist, and consider that they won't continue to exist in the future. It has given parents a new degree of freedom to elicit biological control over their reproduction, and many choose to have babies that do not have Down syndrome. Every potential mother now has a choice to decide if a life is worth living. Sometimes debates around such issues talk about the burden on society, or on families, that people with disabilities pose, or that choosing to have a baby with a particular disability increases the sum total of human suffering. I don't think attempts to apply metrics help here at all. We could equally account for other life choices in terms of burden, such as skiing, or other hobbies or occupations that cause costly injuries.

I'm not advocating a particular view of this incredibly hard question. It's an irreducibly difficult and human one, and although science can furnish us with a context, it doesn't provide a means for making a decision about the value of something as rich as a human life. It is clear that today, we should all be better informed about what people with particular disabilities are like, and their role in society and culture. Informed, compassionate choice should be enshrined in these extraordinarily hard decisions.

CAN YOU FARM HUMANS?

Since the very birth of eugenics, humans have been likened to livestock. Galton did it, Davenport did too, Churchill and pretty

much every eugenics enthusiast observed that animals are bred
to enhance characteristics that we find desirable, and that those
principles should be extended to us. This was often said casually, but also formally in the scientific analyses that bolstered
the eugenics movement. Recall that line from Ronald Fisher
in his classic evolutionary textbook/weird eugenics polemic
The Genetical Theory of Natural Selection: "The deductions
respecting Man are strictly inseparable from the more general
chapters."

Are they inseparable? That was written in 1930, so a few
decades had passed since Galton's first insistence that eugenics was not fundamentally different from agricultural breeding.
Ninety years later, Richard Dawkins, a man whose scientific
work was heavily based on Fisher's (and plenty of other mas-
terminds of mid-twentieth-century evolutionary genetics)
tweeted: "It [eugenics] works for cows, horses, pigs, dogs and
roses. Why on Earth wouldn't it work for humans?"*

It is a good question. The first thing to ask is what does Dawkins mean by "work"? Though this might sound like a bit of
navel-gazing, semantic pseudophilosophy, it is in fact central to
the whole idea of eugenics. All eugenicists in the late nineteenth
and early twentieth centuries confidently asserted the farming
comparison, but without qualifying what it might mean practi-

---

* Though it should be stressed that in a following tweet he pointed out quite
explicitly that taking this view does not in any way endorse eugenics. Furthermore, though Dawkins opposed the removal of Fisher's commemora-
tive window from Gonville and Caius at Cambridge University, he has to my
knowledge never expressed any support for Fisher's political views.

cally. Animal breeding comes in many forms, and farmers have different practices, some of which are less sustainable than others, some of which maximize short-term gain, some of which try to balance the overall health of the flock or herd against the commercial need for them to have market value. Farming practices are wildly different around the world, and the industrialization of farming in the late twentieth century resulted in the creation of vast factories crammed with animals that have huge monetary value, but are deeply unhealthy—chickens that cannot stand, cattle that are beset by mastitis, pigs that are in constant need of medication. There are many farmers who strive to nurture healthy animals and sustainable practices that also have commercial value. Genetics plays a major role in modern agriculture, and often the scientific principles of population genetics are expertly applied to livestock without the farmers necessarily knowing the theoretical framework that underlies it. Just as we consumers don't think too hard about what actually goes into a sausage, we can tend to have a somewhat romantic view of farming, one that is oblivious of the fact that breeding is in part a science but one wedded to the unruly nature of practical biology. Animals have diseases, sometimes as a result of their breeding, and they die or are killed to prevent their genes or susceptibility from contaminating subsequent generations.

Eugenics concerned removing weakness, enhancing desirable traits, both generically and specifically: humans were to be bred to be stronger and smarter, or to eradicate diseases or behaviors. Farming has different aims. You can maximize specific traits, remove weakness, but a healthy, skinny, smart sheep is of no value. Breeding animals to enhance specific traits certainly is done, but if done carelessly, can also result in an overall reduction in the general health of the population. Genes do multiple things; selecting for a gene in one biological context—say, big juicy legs—may well mean also selecting for something less desirable, such as vulnerability to disease. That might yield a more lucrative but more wasteful process, and possibly one that challenges standards of animal welfare—for example, large chickens that are unable to walk, or sheep that are muscular but are prone to birthing problems. These are known issues for farmers, for their livelihood is dependent on the balance of animals that have commercial value and the potential to breed and therefore maintain a stock.

In the United Kingdom in the 1990s, viewers were shocked by news footage of huge piles of sheep being burned, after widespread outbreaks of the disease scrapie. Only a few years before, European cow populations had been shredded by BSE (or mad cow disease), and this had transmitted to humans with lethal consequences. Scrapie is the ovine version of BSE and does not transmit to humans. While it is present at low levels in black-faced sheep in the United States, in Europe it is still a terrible disease that decimates flocks and has a major impact on sheep farming. Like BSE and its human counterpart Creutzfeldt-Jakob disease, lumps of proteins aggregate in the brains of sheep with scrapie and cause holes to develop, resulting in neurological damage and eventually death. There is no treatment. Different breeds have vastly different susceptibilities to scrapie, Cheviots and Suffolks being the most vulnerable, and the genetics that underlie susceptibility was worked out, such that shepherds could breed out this weakness. In 2001, the British government issued the National Scrapie Plan, which included a five-point grading system (from R1 through R5) for susceptibility in individual sheep. This was based on the precise genetics of the protein that causes scrapie being characterized, just as we have done for every genetic disease in humans. R1 sheep are the most genetically resistant, and R5 the least. Farmers were instructed to breed only R1s and 2s. However, via a costly process of trial and error, what farmers found is that flocks bred from R1s didn't get scrapie but they did suffer from various other conditions, including lameness and mastitis. Which also meant the flocks were not viable. Instead, careful programs of breeding with the R2 and R3 sheep resulted in a general resistance to scrapie, and a generally healthier stock. Some smaller farms are beginning to adopt a method called mob breeding, where instead of focusing a breeding population on enhancing the very best or biggest sheep, the farmers cull only the weakest. The idea is that this is a more sustainable form of farming that results in stable and commercially viable flocks, and a degree of outbreeding and genetic diversity that should keep sheep healthy. But it still requires killing the weak. What is the acceptable level of culling "weak" humans to maintain an overall healthy population?

The universal agricultural analogy is poor. Animals are mutable, and humans are animals, but beyond that, comparing humans with agriculture is fundamentally flawed. The rose has been bred to be radically different from its ancient natural ancestor, but in doing so it requires constant care and special soils and food to maintain its unnatural beauty. We have carefully bred dogs to have a phenomenal range of characteristics, which means sacrificing other traits, such that a whippet is quicksilver fast but won't retrieve, and a retriever retrieves but is a useless ratter, and a Bichon Frise is cute but doesn't do much else. Furthermore, the traits we have bred into dogs for both practical and aesthetic reasons have also resulted in serious problems that are arguably cruel: pugs that cannot breathe properly, bloodhounds whose droopy eyes are prone to infection, dachshunds who suffer spinal disc pain.

The cows that provide us with milk have been bred to do so only in the confines of a farm. Many sheep breeds need to live and reproduce indoors. Farm breeding is not concerned with making animals generally better or brighter, but more precisely designed for the farm itself. In the wild, most farm animals would not fare so well.

The artificial selection of farming is also concerned with uniformity. There is another form of biological control that we should also consider because its commitment to uniformity is supreme. Genetic intervention into farm animals took a giant leap two decades ago. Some of you will recall the most famous sheep of the twentieth century. In July 1997, the world was introduced to Dolly. Born in February 1996, Dolly was the first cloned mammal—though this is a record that requires a touch of clarification. Dolly wasn't the first cloned mammal, nor even the first cloned sheep. That honor went to an unnamed sheep, who had been created from embryonic cells by the Danish scientist Steen Willadsen in 1984.

Dolly was the first cloned sheep from an adult somatic cell who lived. What this means is that the nucleus of an adult cell* from a Finn Dorset ewe had been removed and inserted into the unfertilized egg of a second ewe. The nucleus contains (almost) all of the genetic material for a mammal, and therefore every gene required to take an egg from conception to death. For the previous several thousand years, sheep and their ovine ancestors had acquired a full set of genes by harboring half a set in an egg, and completing it with another half contained within a lucky sperm. Dolly's genome was already complete before being inserted into an empty egg, which was treated with a small electric charge to stimulate cell division, then implanted into a surrogate mother who carried her to term. It took three sheep to make one, which is less efficient than nature's method, but it was a huge scientific achievement nonetheless.

However, it didn't really take three sheep at all. Dolly was the 277th attempt, twenty-nine of which resulted in viable embryos, three of which reached term, and only Dolly lived. The success rate was about 0.4 percent. Six years later, Prometea was born in Italy, the first cloned horse, the only successful live birth out of 814 attempts.† The mechanics of cloning and indeed any manipulation of a biology that is billions of years old is not that easy to control, even with all our skills, tools and knowledge.

---

* The source of Dolly's entire genetic material was taken from a mammary cell. She was named after Dolly Parton because she has large breasts. I am truly embarrassed on behalf of my profession.

† Since then, a few horses have been successfully cloned and studied primarily for the racing industry.

Cloning technology has considerably moved on, and this technique is not the same as genetic engineering. Dolly's true scientific legacy was to help develop new methods in stem-cell research and more generally a richer understanding of the biology of very early life. Cloning is not closely tied to breeding nor eugenics, and I don't think it will have a directly significant role in the future of farming, though in China there is now a major facility thought to be dedicated to cloning farm animals. I bring up Dolly really to show how inefficient a process it was. Many of you reading this will be vegetarians or vegans who morally object to any animal farming on the grounds that no matter how high the standards of animal welfare is in those farms, killing them for us is unacceptable. There is a huge range of standards in breeding and welfare in the millions of farms around the world, but in all cases we exert an almost absolute tyrannical control over the animals' sex lives in order to produce the meat that we eat and milk that we drink. And still the process is inefficient and experimental and comes with unforeseen consequences, which we as consumers are largely oblivious of. Humans can be bred, but for this to be a valid argument in defense of eugenics, you're going to have to either radically lower the welfare with which we indulge people or radically improve the lives of sheep.

CONTROL

The sentiment "Of course, eugenics would work" really only means one thing: we are evolved. It is blandly true and also meaningless. The principles of genetics are universal; the cod-

ing system used to go from DNA to proteins to cells to a creature is the same too. Sexual reproduction is mostly the same,
but there are plenty of animals for whom it is not. All those
things are true, and all life is on the same Darwinian tree,
evolved from the same set of deep ancestors from whom our
fundamental biology is derived. But we're not the same. We're
not mice, nor cells in a petri dish. We're not farm animals either,
bred over thousands of generations specifically for traits that we
find desirable, often at the cost of independence or autonomy,
and sometimes health.

The attempted purification of humankind in the policies of
the Third Reich failed and was always going to fail. These policies were unhinged and cruel, but at least based on what their
proponents believed was a science. However, they were wrong
and wrongheaded. They were bewitched by a nascent science
that was decades away from being even partially understood.
The progress we have made in genetics since then has been spec-
tacular, and in the last few years the pace has accelerated mag-
nificently. But it is my contention that with our knowledge of
human genetics as it currently is, the eugenic visions of Galton,
Ploetz and Davenport are not much more realistic than they
were a century ago. The complexities revealed in the throng-
ing networks of endless genetic variants in human beings make
selection for many of the traits targeted in the past *harder* to
understand than in those pre-molecular days. Yes, we can breed
humans and yes, we are replete with heritable traits. If we were
to enact eugenics policies today, would we really know what we
are selecting for or against?

You could make the argument that had the eugenics policies

enacted in the United States and Germany (and proposed in the United Kingdom) been enforced with today's knowledge and tools, the world would've been robbed of Woody Guthrie, who died of Huntington's. Perhaps the embryo that became the actor and comedian Robin Williams, who was diagnosed with Parkinson's, ADHD and bipolar diseases, would have been screened out of existence before he was born, had that tech been available. Maybe we would've had a completely different Princess Leia, Sarah Connor and Scarlett O'Hara, if the actors Carrie Fisher's, Linda Hamilton's and Vivien Leigh's bipolar had been spotted when they were embryos. Twenty genes have been identified that characterize the risk of the motor neurone disease that Stephen Hawking had. Would the world rather that he had not existed?

Maybe we might wander beyond the realm of known diagnoses and come to the realization that some of the more colorful characteristics of much loved historical figures were actually undiagnosed conditions that an alternate eugenic timeline might have precluded their being. Maybe Beethoven would never have existed, perhaps not because of his deafness as J. B. S. Haldane speculated, but because he suffered from depression and mania throughout his troubled life, which some scholars speculate may well have been bipolar disorder. Or Isaac Newton, whom you could classify as simply very odd due to his persistently peculiar behaviors, but may well have been autistic. Newton endured fits of mania and depression too, and some have plausibly considered that he had schizophrenia. Abraham Lincoln also suffered from profound depressions, and some scientists and doctors have speculated that he had a rare genetic disorder, maybe Marfan syndrome, or possibly multiple endocrine neoplasia type 2B, which typically results in people being tall and lanky, with an elongated face.

The significant depression and alcoholism that defined Winston Churchill's life almost certainly will have had an underlying genetic architecture that would be identifiable in this era of endless genomic data. How ironic that he put so much effort into the eradication of disabilities in others. Or perhaps even the father of eugenics himself, Francis Galton: an obsessive data collector, an ultrasystemizing brain, with limited empathy, who expressed little interest in sex and died heirless. Maybe he had some diagnosable psychiatric disorder too; maybe he was also gay, or asexual, and destined never to commit his genes into future generations. Maybe those characteristics were encoded in his DNA. What would have become of those genotypes, if instead of being housed in the zygotes that became Churchill or Galton, they were born in a workhouse in Hackney, or in abject poverty in turn-of-the-century California, or in Berlin in 1933?

Or perhaps we should not bother indulging a post hoc alternative reality where these great people were never allowed to exist, and instead remember that mental illnesses affect one in four people during their lives, and yet those people have no less value than those Galtonian men of eminence. We fetishize mental health issues as the source of greatness when they are in the great, and curse them when it is everyone else. Perhaps instead we should recognize that those criteria that sat in the foundations of eugenics were not absolute metrics of human worth, but were often arbitrary value judgments issued by decree of the powerful. Undesirable, defective, disabled—these are political terms, which change with time, whim and culture. Perhaps we should focus our cultural attention not on the people who are great, but on those who are merely good.

Eugenics began as an idea intended for the improvement of a people. But you cannot have desirable characteristics without an implicit hierarchy that some other characteristics are undesirable. Consider that, hypothetically, when you select those hundreds of slight genetic variants for intelligence, you may well be selecting for or against other traits as well. You may be selecting against compassion, or fertility, or another behavior that is desirable or defective. I don't know. But neither do you, nor does anyone else.

The casualness with which politicians, journalists and even some scientists confidently assert what they think is a fact is striking. That rings true in the past and today.

––––––––––

Those who ponder a new eugenics, or embryo selection for things like intelligence, or set up companies to sell this service, are careless people—as *The Great Gatsby*'s narrator says— people who say things easily and let others clear up the mess they make. Scientists cannot afford to be careless, and no one can afford to be careless when creating new people.

The control we exert over biology today is astonishing, but it remains haphazard. It will only get better, as we continue to explore our genomes, as we invent new techniques to tweak, prod and poke at our evolved bodies. I have no doubt that these advances will continue to take us forward toward a greater understanding of fundamental biology and of how to treat diseases such as the cancers that blight our lives but will one day be of interest only to historians. I also have no doubt that these advances will propel the possibility of eugenic-style modification and selection of our children.

Nevertheless, I have spent much of the last twenty-five years thinking, writing and working with human genetics, and I cannot see a way in which genetic modification to improve the stock of people could be achieved without costs that dwarf arguably marginal benefits. As for embryo selection, the tech is already in place, immature though it is, but I'd be skeptical about its utility for simple traits such as eye color, and cynical about it for much more complex traits, such as for behaviors or intelligence, or to curtail the risk of mental health problems. I cannot see how this is a plausible—or desirable—future.

More than eighty years after Haldane said it, his words are still true: "I do not believe our present knowledge of human heredity justifies such steps."

If you wish to apply metrics, status and ranks to people, such that we might shift society toward improvement, consider this: none of the worst crimes of humanity—the genocides, war crimes, chemical warfare, ethnic cleansing, financial crashes and environmental desecration, invasions, colonizations, rapes and murders—not one of these was perpetrated by people with Down syndrome. No one with achondroplasia has ever committed mass murder or propped up a genocidal campaign.*

Sometimes, proposals for population control are couched in utilitarian terms: actions justified by the metric of the greatest success or happiness for the greatest number of people. But there is nothing to indicate that the eradication of people with below-average IQs (by definition, half of all people), or with physical disabilities, or from minority groups, or simply from families locked into cycles of poverty, would increase the sum total of human happiness, when those people are not responsible for the acts that have caused the most suffering. If you wish, as the eugenicists did, to reduce people to metrics, then this surely is a simple logic. Is this not the real utilitarian argument? If we truly wanted to reduce the sum total of human suffering then we should eradicate the powerful, for wars are fought by people but started by leaders.

Are the great people, the achievers, the women and men who change history, the countries who define the global stage, are they genetically predisposed to their greatness, or is it the cosmic happenstance of birth? The answer, in that most annoying way that scientists often rely on, is a bit of both. All characteristics are heritable. People are not born equal. Our genes differ, our environments differ. We are shackled to both. None of these

* It has been pointed out to me that Tyrion Lannister did exactly this, but I've never seen *Game of Thrones*, and also it's not real.

differences stratify by clear lines that would warrant meddling in the crude engineering manner of the Edwardian eugenicists. Nor do they reveal sharp distinctions today that can be isolated in our DNA, such that we could select them with any accuracy.

Am I better than you? Am I smarter than you? Are my genes better than yours? If any of these questions could possibly be answered with a yes, do I therefore deserve to preserve my genes with priority over others? Would the preservation and multiplication of those genes result in a better society? I think the answer to that is unknowable, but I suspect that it is no. I do not think I am better than you. I think all people have value, and a good society is one that protects its most vulnerable members, rather than erasing them. Nowhere in the conversations we have about improvement or betterment of society do we focus on traits such as compassion or kindness. Why not? Because eugenics and the policies surrounding it are issued only by the powerful to maintain control.

Eugenics is a busted flush, a pseudoscience that cannot deliver on its promise. Maybe that will change in time, as we anatomize our genomes ever more precisely. Maybe we will uncover the signals of traits in our DNA with such sophistication that the latest embryo selection tools and the newest gene editing techniques will be suitably deployed to enhance them, with only benefits to the individual. Or maybe we won't, and will find that the trajectory of our understanding of genetics is on the right path, and there are no revelations to come, only details that will further preclude meddling.

And besides, as with all technology, these techniques are expensive, and only available to the rich—individuals, particular strata of society, and the wealthiest of countries. At its inception, eugenics was the fetishization of a science for political ends. It is no different today. Instead of grasping for science at the edges of our understanding, would it not be better to improve a people with mechanisms that work? Via education, health care and equality of opportunities regardless of family history or the luck of the draw.

Ultimately, we are considering risk. The predictions we can make about biology are sound but far from perfect. It is normal to wish to reduce the chance of danger or disease, and elicit control, especially over the things that we love more than anything else: our children, our families and our creed.

Genetics is the most awesome force the planet has ever seen, but we wield it like a kid who's found their dad's gun. No one wants to increase the sum total of human unhappiness, unless they are determinedly cruel. Everyone wants their children to thrive, to live with minimal pain and with fair and equal opportunities. We all want our kin, our tribe, our friends and compatriots to succeed in their endeavors, in their pursuit of life, liberty and happiness. But what are we willing to do to ensure it?

ACKNOWLEDGMENTS

As always, I rely on a cadre of generous experts, consisting of scientists, historians, writers and a shepherd, to advise me and help structure my thoughts and words. Any errors are entirely my own.

Six people have gone to extraordinary lengths to thoughtfully and patiently panel-beat the science, history and my ideas into better shape, and they really didn't have to. I am particularly indebted, yet again, to Joe Cain, Matthew Cobb, Peter Frankopan, Ana Paula Lloyd, Alice Roberts and Mark Thomas.

I have also received significant help and support from many friends and colleagues. The work of each is of the highest quality and I urge you to seek them out: Abdel Abdellaoui, the Authors XI Cricket Club, Daniel Benjamin, Ewan Birney, Sarah Churchwell, Mark Compton-James, George Davey Smith, Caroline Dodds Pennock, Alexandra Fair, Alan Fersht, Simon Fisher, Nishani Frazier, Hannah Fry, Stephen Fry, Suzi Gage, Alex Garland, Kathryn Paige Harden, Natalie Haynes, Tom Holland, Turi King, Omar Little, Elspeth Merry Price, Shaun O'Boyle, David Olusoga, Helen O'Neill, Radiohead, Jennifer Raff, Nina Ramirez, Lynsey Reader, James Rebanks, Fern Riddell, Stuart Ritchie, Ananda Rutherford, Aylwyn Scally, Brandy Schillace, Katherine Schofield, Tom Shakespeare, Jane Sowden, Dallas Swallow, Matthew Sweet, Michael Taylor, Nicola Tomlinson, Veronica van Heyningen, Robert Winston and Victoria Woodham.

The people that enable me to write are the finest of friends and family. Georgia, Bea, Jake and Juno are my lights and rocks. Thanks to my long-standing friend, collaborator and agent Will Francis, and the team at Janklow & Nesbit, and to my U.K. editor Jenny Lord. For this, the U.S. version, my immense gratitude goes to Trent Duffy, Rebecca Homiski and above all Tom Mayer, who immeasurably helped make sense of this immensely complex story.

REFERENCES

PART ONE: QUALITY CONTROL

Baker, G. J. 2014. "Christianity and Eugenics: The Place of Religion in the British Eugenics Education Society and the American Eugenics Society, c. 1907–1940." *Social History of Medicine* 272: 281–302.

Barlow, T., and Armit H. W. Ilkeston. 1911. "International Hygiene Exhibition, Dresden, 1911." *British Medical Journal* 1 (January 28): 225.

Blacker, C. P. 1952. "J. B. S. Haldane on Eugenics." *Eugenics Review* 44 (3): 146–51.

Bodmer, W. F. 2017. "A Haldane Perspective from a Fisher Student." *Journal of Genetics* 96 (5) (November): 743–46.

Boudreau, E. B. 2005. "'Yea, I Have a Goodly Heritage': Health Versus Heredity in the Fitter Family Contests, 1920–1928." *Journal of Family History* 30 (4) (October): 366–87.

Brown, S. 2009. "Hitler's Bible: An Analysis of the Relationship Between American and German Eugenics in Pre-war Nazi Germany." *Vesalius* 15 (1) (June): 26–31.

Chapman, M. 1980. "Infanticide and Fertility Among Eskimos: A Computer Simulation." *American Journal of Physical Anthropology* 53 (2) (August): 317–27.

Cowlishaw, G. 1978. "Infanticide in Aboriginal Australia." *Oceania* 48 (4): 262–83.

Davenport, C. B., and E. B. Muncey. 1916. "Huntington's Chorea in Relation to Heredity and Eugenics." *American Journal of Insanity* 73 (2): 195–222.

Dejong-Lambert, W. 2017. "J. B. S. Haldane and ЛысеНкОвщиНа (*Lysenkovschina*)." *Journal of Genetics* 96 (5): 837–44.

Edwards, A. W. 2000. "The Genetical Theory of Natural Selection." *Genetics* 154 (4): 1419–26.

Fisher, R. A. 1915. "The Evolution of Sexual Preference." *Eugenics Review* 7 (3): 184–92.

Gökyiğit, Emel Aileen. 1994. "The Reception of Francis Galton's 'Hereditary Genius' in the Victorian Periodical Press." *Journal of the History of Biology* 27 (2): 215–40.

Grodin, M. A., E. L. Miller and J. I. Kelly. 2018. "The Nazi Physicians as Leaders in Eugenics and 'Euthanasia': Lessons for Today." *American Journal of Public Health* 108 (1): 53–57.

"H." 1869. "Darwinism and National Life." *Nature* 1: 183–84.

Inge, W. R. 1909. "Some Moral Aspects of Eugenics." *Eugenics Review* 1 (1): 26–36.

Karp, R. J., Q. H. Qazi, K. A. Moller et al. 1995. "Fetal Alcohol Syndrome at the Turn of the 20th Century: An Unexpected Explanation of the Kallikak Family." *Archives of Pediatrics and Adolescent Medicine* 149 (1) (January): 45–48.

Kruse, Horst H. 2015. "F. Scott Fitzgerald and Mary Harriman Rumsey: An Untold Story." *F. Scott Fitzgerald Review* 13 (1): 146–62.

Lawing, Sean B. 2013. "The Place of the Evil: Infant Abandonment in Old Norse Society." *Scandinavian Studies* 85 (2): 133–50.

Lee, R. B. 1980. "Lactation, Ovulation, Infanticide and Women's Work." In

Biosocial Mechanisms in Population Regulation, edited by M. Cohen et al., 321–48. New Haven: Yale University Press.

Lovett, L. L. 2007. "Fitter Families for Future Firesides: Florence Sherbon and Popular Eugenics." *Public Historian* 29 (3) (Summer): 69–85.

Pearson, Karl. 1900. *National Life from the Standpoint of Science*. London: Adam and Charles Black.

Proctor, R. J. 1988. *Racial Hygiene: Medicine Under the Nazis*. Cambridge, Mass.: Harvard University Press.

Reilly, Philip R. 1987. "Involuntary Sterilization in the United States: A Surgical Solution." *Quarterly Review of Biology* 62 (2) (June): 153–70.

"The Report of the Royal Commission on the FeebleMinded." 1908. *British Medical Journal* 2 (2485): 415–18.

Rosenberg, Charles E. 1961. "Charles Benedict Davenport and the Beginning of Human Genetics." *Bulletin of the History of Medicine* 35 (3) (May–June): 266–76.

Schuller, K. 2018. "From Impressibility to Interactionism: W. E. B. DuBois, Black Eugenics, and the Struggle Against Genetic Determinisms." In *The Biopolitics of Feeling: Race, Sex, and Science in the Nineteenth Century*, 172–204. Durham, N.C.: Duke University Press.

Selden, S. 2005. "Transforming Better Babies into Fitter Families: Archival Resources and the History of the American Eugenics Movement, 1908–1930." *Proceedings of the American Philosophical Society* 149 (2) (June): 199–225.

Smith, J. D., and M. L. Wehmeyer. 2012. "Who Was Deborah Kallikak?" *Intellectual and Developmental Disabilities* 50 (2): 169–78.

Sparkes, R. 1999. "The Enemy of Eugenics." *Chesterton Review* 25: 117–29.

Taeuber, I. B. 1938. "Haldane Looks at Eugenics: A Review of *Heredity and Politics*." *Journal of Heredity* 29 (8) (August): 304–5.

Wallace, A. 1870. "Hereditary Genius: An Inquiry into Its Laws and Consequences." *Nature* 1: 501–3.

Weisbord, Robert G. 1973. "Birth Control and the Black American: A Matter of Genocide?" *Demography* 10 (4): 571–90.

Weiss, Sheila Faith. 1987. "The Race Hygiene Movement in Germany." *Osiris* 3: 193–236.

Willadsen, S. 1986. "Nuclear Transplantation in Sheep Embryos." *Nature* 320: 63–65.

PART TWO: SAME AS IT EVER WAS

Agar, N. 2019. "Why We Should Defend Gene Editing as Eugenics." *Cambridge Quarterly of Healthcare Ethics* 28 (1): 9–19.

Anomaly, J. 2018. "Defending Eugenics." *Monash Bioethics Review* 35: 24–35.

Bayefsky, Michelle J. 2016. "Comparative Preimplantation Genetic Diagnosis Policy in Europe and the USA and Its Implications for Reproductive Tourism," *Reproductive Biomedicine & Society Online* 3: 41–47.

Chabris, C. F., J. J. Lee, D. Cesarini et al. 2015. "The Fourth Law of Behavior Genetics." *Current Directions in Psychological Science* 24 (4): 304–12.

Claussnitzer, M., J. H. Cho, R. Collins et al. 2020. "A Brief History of Human Disease Genetics," *Nature* 577: 179–89.

Davies, G., M. Lam, S. E. Harris et al. 2018. "Study of 300,486 Individuals Identifies 148 Independent Genetic Loci Influencing General Cognitive Function." *Nature Communications* 9 (1): 2098.

Dennison, C. A., S. E. Legge, A. F. Pardiñas et al. 2020. "Genome-Wide Association Studies in Schizophrenia: Recent Advances, Challenges and Future Perspective." *Schizophrenia Research* 217 (March): 4–12.

Edwards, R., and R. Gardner. 1967. "Sexing of Live Rabbit Blastocysts." *Nature* 214: 576–77.

Gershon, E. S. 1997. "Ernst Rüdin, a Nazi Psychiatrist and Geneticist." *American Journal of Medical Genetics* 74 (4) (July 25): 457–58; author reply 461–63.

Ghosh, S., E. Feingold and S. K. Dey. 2009. "Etiology of Down Syndrome: Evidence for Consistent Association Among Altered Meiotic Recombination, Nondisjunction, and Maternal Age Across Populations." *American Journal of Medical Genetics Part A* 149A (7) (July): 1415–20.

Handyside, A., E. Kontogianni, K. Hardy et al. 1990. "Pregnancies from Biopsied Human Preimplantation Embryos Sexed by Y-Specific DNA Amplification." *Nature* 344: 768–70.

Joseph, J. 2001. "Separated Twins and the Genetics of Personality Differences: A Critique." *American Journal of Psychology* 114 (1) (Spring): 1–30.

Joseph, J., and N. A. Wetzel. 2013. "Ernst Rüdin: Hitler's Racial Hygiene Mastermind." *Journal of the History of Biology* 46 (1) (Spring): 1–30.

Karavani, Ehud, Or Zuk, Danny Zeevi et al. 2019. "Screening Human Embryos for Polygenic Traits Has Limited Utility," *Cell* 179 (6): 1424–35.e8.

Karpinski, R. I., A. M. K. Kolb, N. A. Tetreault et al. 2018. "High Intelligence: A Risk Factor for Psychological and Physiological Overexcitabilities." *Intelligence* 66 (January–February): 8–23.

Kranzler, H. R., H. Zhou, R. L. Kember et al. 2019. "Genome-Wide Association Study of Alcohol Consumption and Use Disorder in 274,424 Individuals from Multiple Populations." *Nature Communications* 10 (1): 1499.

Liang, P., Y. Xu, X. Zhang et al. 2015 "CRISPR/Cas9-Mediated Gene Editing in Human Tripronuclear Zygotes," *Protein Cell* 6 (5): 363–72.

Ma, H., N. Marti-Gutierrez, S. W. Park et al. 2017. "Correction of a Pathogenic Gene Mutation in Human Embryos." *Nature* 548: 413–19.

Mai, C. T., J. L. Isenburg, M. A. Canfield et al. 2019. "National Population-Based Estimates for Major Birth Defects, 2010–2014." *Birth Defects Research* 111 (18): 1420–35.

Morgan, T. H. 1925. *Evolution and Genetics,* 2nd ed. Princeton, N.J.: Princeton University Press.

Savage, J. E., P. R. Jansen, S. Stringer et al. 2018. "Genome-Wide Association Meta-Analysis in 269,867 Individuals Identifies New Genetic and Functional Links to Intelligence." *Nature Genetics* 50: 912–19.

Sharma, H. D. Singh, A. Mahant et al. 2020. "Development of Mitochondrial Replacement Therapy: A Review." *Heliyon* 6 (9) (September 14).

Susser, E. S., and S. P. Lin. 1992. "Schizophrenia After Prenatal Exposure to the Dutch Hunger Winter of 1944–1945." *Archives of General Psychiatry* 49 (12) (December): 983–88.

Susser, E., R. Neugebauer, H. W. Hoek et al. 1996. "Schizophrenia After Prenatal Famine: Further Evidence." *Archives of General Psychiatry* 53 (1) (January): 25–31.

Tang, L., Y. Zeng, H. Du et al. 2017. "CRISPR/Cas9-Mediated Gene Editing in Human Zygotes Using Cas9 Protein." *Molecular Genetics and Genomics* 292 (3): 525–33.

Torrey, E. F., and R. H. Yolken. 2010. "Psychiatric Genocide: Nazi Attempts to Eradicate Schizophrenia." *Schizophrenia Bulletin* 36 (1): 26–32.

Turkheimer, E. 2000. "Three Laws of Behavior Genetics and What They Mean." *Current Directions in Psychological Science* 9 (5): 160–64.

Veit, W., J. Anomaly, N. Agar et al. 2021. "Can 'Eugenics' Be Defended?" *Monash Bioethics Review* 39: 60–67.

Wilson, Kalpana. 2017. "In the Name of Reproductive Rights: Race, Neoliberalism and the Embodied Violence of Population Policies." *New Formations* 91: 50–68.

Wilson, R. A. 2019. "Eugenics Undefended." *Monash Bioethics Review* 37: 68–75.

Zeng, Y., J. Li, G. Li et al. 2018. "Correction of the Marfan Syndrome Pathogenic FBN1 Mutation by Base Editing in Human Cells and Heterozygous Embryos." *Molecular Therapy* 26 (11) (November 7): 2631–37.

INDEX

feminists, 53

fetal alcohol spectrum disorder, 91

FGFR3, 155–56

First International Eugenics Congress, 85

First Nations, 12, 148–49

Fischer, Eugen, 119, 121, 124

Fisher, Ronald Aylmer, 15, 30–31, 95–97, 102–3, 132, 135. See also *The Genetical Theory of Natural Selection*

Fisherian runaway selection, 98

Fitter Family competitions, 69–70, 72–73

Fitzgerald, F. Scott, 65–66

forced sterilizations

advocated by Jordan, 16

African Americans, 74

authorized by Wilson, 15

conditions for in Nazi Germany, 122

as continuing, 147–49

Davenport's views, 63

Immigration and Customs Enforcement detention centers, 12

as legal, 63, 70–71

Mental Deficiency Bill/Act and, 86

of mixed-race people, 125

number in US in twentieth century, 71

poverty and, 86

rejection of, 140–41

Sharp's views, 77

Ford, Henry, 125, 134

Frankenstein (Shelley), 3, 4

freedom, 7, 8, 141

Galton, Francis

overview, 37–44

as childless, 85

eugenics and statistics, 98–99

financial endowment, 134–35

founding Eugenics Records Office, 14

genius work, 39–48

influence of, 48–54, 56, 62, 92–93, 130–31

as obsessive, 235

Pearson and, 93–94

religion and, 82

science replacing religion, 57

twin studies, 159

UCL removing name, 131

white supremacy and, 57

writing novel, 57–58

See also *Hereditary Genius*

Galton Laboratory for National Eugenics, 14–15, 92, 95, 106, 131

Gandhi, Indira, 149–50

Gandhi, Sanjay, 150